Nanotechnology in the Life Sciences

Series Editor

Ram Prasad
Department of Botany
Mahatma Gandhi Central University
Motihari, Bihar, India

Nano and biotechnology are two of the 21st century's most promising technologies. Nanotechnology is demarcated as the design, development, and application of materials and devices whose least functional make up is on a nanometer scale (1 to 100 nm). Meanwhile, biotechnology deals with metabolic and other physiological developments of biological subjects including microorganisms. These microbial processes have opened up new opportunities to explore novel applications, for example, the biosynthesis of metal nanomaterials, with the implication that these two technologies (i.e., thus nanobiotechnology) can play a vital role in developing and executing many valuable tools in the study of life. Nanotechnology is very diverse, ranging from extensions of conventional device physics to completely new approaches based upon molecular self-assembly, from developing new materials with dimensions on the nanoscale, to investigating whether we can directly control matters on/in the atomic scale level. This idea entails its application to diverse fields of science such as plant biology, organic chemistry, agriculture, the food industry, and more.

Nanobiotechnology offers a wide range of uses in medicine, agriculture, and the environment. Many diseases that do not have cures today may be cured by nanotechnology in the future. Use of nanotechnology in medical therapeutics needs adequate evaluation of its risk and safety factors. Scientists who are against the use of nanotechnology also agree that advancement in nanotechnology should continue because this field promises great benefits, but testing should be carried out to ensure its safety in people. It is possible that nanomedicine in the future will play a crucial role in the treatment of human and plant diseases, and also in the enhancement of normal human physiology and plant systems, respectively. If everything proceeds as expected, nanobiotechnology will, one day, become an inevitable part of our everyday life and will help save many lives.

More information about this series at http://www.springer.com/series/15921

Juan Bueno

Preclinical Evaluation of Antimicrobial Nanodrugs

Juan Bueno
Senior Researcher, Research Center of Bioprospecting and Biotechnology
for Biodiversity Foundation (BIOLABB)
Armenia, Quindio, Colombia

ISSN 2523-8027 ISSN 2523-8035 (electronic)
Nanotechnology in the Life Sciences
ISBN 978-3-030-43857-9 ISBN 978-3-030-43855-5 (eBook)
https://doi.org/10.1007/978-3-030-43855-5

This Springer imprint is published by the registered company Springer Nature Switzerland AG
The registered company address is: Gewerbestrasse 11, 6330 Cham, Switzerland

Dedication

Dedicated to

Martha Ritoré for her full support and management for the realization of this book

Sebastian Ritoré for his constant work to improve on the road that awaits us

ITC Ritoré for her eternal and unconditional company during the writing of this manuscript

> "Here we stand, bound forevermore
> We're out of this world, until the end
> Here we are, mighty, glorious
> At The End Of The Rainbow
> With gold in our hands"
> —*Andreas Mueck/Joacim Anders Cans, Martin Albrecht-Lindow*
> "End of the Rainbow" by Hammerfall, Legacy of Kings

Preface

We are facing one of the biggest threats in public health and perhaps one of the biggest medical crises in human history. Resistance to antimicrobial drugs is progressively increasing without a blunt response to their control and treatment. This lack of a scientific and political response puts all communities at risk of being affected by the emergence of new resistant microorganisms as well as the acquisition of resistance mechanisms by those that were already believed to be controlled, taking also into consideration the possibility of a global transmission due to the increase in transoceanic travel. Thus, in this order of ideas, a revolutionary innovation is required in the design and development of new antimicrobial drugs that allow greater action within the framework of a personalized medicine and avoid the emergence of resistance with low toxicity in patients affected by infectious diseases. In this way, nanotechnology represents this advance of impact in the advent of a modern anti-infective pharmacology. Thus, it is possible using this technology to obtain new formulations with nanocarriers and develop new antibiotics combined with nanoparticles and nanostructure design that scientists can produce both nanosurfaces and antimicrobial nanocoatings capable of reducing the transmission of infectious disease from pandemic outbreaks. But it is also important to remember that in a modern symbiotic medicine where the holobiont (host and its symbiotes), the following guidelines should be taken into account for the development of new antimicrobials:

- Anti-infectives should have the highest possible affinity for the target microorganism.
- They should have pharmacokinetic and pharmacodynamic parameters capable of being adapted to the pharmacogenetic aspects of the patient in a personalized medicine model.
- They should comply with Lipinski's rules of five to a greater degree.
- In this way, modern antimicrobials should not alter the host microbiome and not generate the phenomenon of dysbiosis.

- To a greater extent, symbiotic anti-infective therapy should seek to restore equilibrium interaction between the host and its symbionts as part of the healing process.
- Finally, the possible toxicities and adverse reactions must be clearly determined in order to administer the new medications safely and adequately.

Thus, the promise of nanotechnology to initiate a pharmacological revolution requires the redesign of innovative preclinical evaluation models that take into account new approaches beyond the terms microbicide and microbiostatics such as antibiotic adjuvants, the inhibition of virulence mechanisms and of quorum-sensing system signals, and the antibiofilm activity, as well as preventing adhesion and colonization in the case of nanosurfaces in hospital spaces and on medical equipment. In the same way, it is necessary to establish the correct parameters of nanotoxicity that allow scientists to evaluate the safety and lethal dose of nanomaterials in patients and the environment.

For the above reasons, the main objective of this book is to analyze in the light of the concepts of personalized and symbiotic medicine the preclinical aspects of the evaluation of the antimicrobial activity of nanomaterials in order to establish clear rules of the game with which researchers in antibiotic therapy and the industry achieve a greater impact on the greatest global threat of our times of climate crisis, always with the aim that every living organism in the biosphere seeks survival and persistence.

Quindio, Colombia Juan Bueno

Contents

Chapter 1
Antimicrobial Screening: Foundations and Interpretation

It has been shown to be possible, by deliberately planned and chemotherapeutic approach, to discover curative agents which act specifically and aetiologically against diseases due to protozoal infections, and especially against the spirilloses, and amongst these against syphilis in the first place. Further evidence for the specificity of the action of dihydroxydiaminoarsenobenzene [Salvarsan '606'] is the disappearance of the Wasserman reaction, which reaction must … be regarded as indicative of a reaction of the organism to the constituents of the spirochaetes
—Paul Ehrlich (1854–1915)

A rise in body temperature during sulphonamide treatment intensifies the biochemical reaction between drug and pathogen, while at the same time the heat itself injures the heat-sensitive gonococci.
—Gerhard Domagk (1895–1964)

Abstract It is a fact that the current antimicrobial susceptibility protocols do not belong to an adequate precision personalized medicine model. In this way, these methods can predict the therapy that will fail but not the therapy that will be successful, which limits its range of action and predisposes to the appearance of antimicrobial resistance. Thus, it is necessary to develop comprehensive models of antimicrobial susceptibility that combine antibiotic activity, pharmacokinetics, and virulence factors within the same protocol in order to predict the clinical response to antimicrobial treatment. For this reason, theranostics should be the model to be used in order to develop modern biosensors capable of detecting infectious disease and determining the appropriate treatment with the ability to predict its success and cure. In this order of ideas, the objective of this chapter is to rethink the disadvantages of current susceptibility methods in order to provide comprehensive solutions that allow the development of new methods that can slow the spread of antimicrobial resistance, as well as the development of new anti-infective medications.

J. Bueno, *Preclinical Evaluation of Antimicrobial Nanodrugs*, Nanotechnology in the Life Sciences, https://doi.org/10.1007/978-3-030-43855-5_1

1.1 Introduction

Antimicrobial screening is more than an in vitro evaluation protocol to determine inhibitory activity on certain microorganisms; it is a part of a reproducible system of correlation and prediction of the emergence of drug resistance, so it should be considered a method for making decisions both in the biomedical research and in clinical care (Jenkins and Schuetz 2012; Balouiri et al. 2016; Kim et al. 2019). In this order of ideas, any antimicrobial susceptibility test (AST) should allow the selection of the most potent molecule, substance, formulation, or combination of compounds against a certain infectious agent and that also allows predicting the possible clinical response to treatment under certain circumstances (Li et al. 2017; Kurhekar et al. 2019; Maugeri et al. 2019). But in this sense, the question arises of how an in vitro test has the ability to predict the behavior of a drug inside an infected patient (Menden et al. 2019), as well as the type of biomarker that must be evaluated in order to determine if a microorganism is resistant or susceptible to antimicrobial treatment (Tannert et al. 2019; Volz et al. 2019). Equally very important to take into account is that the AST reading should be highly accurate, with the end to classify the microorganisms as susceptible or resistant based on the patient's condition, so it has become a topic of interest in personalized medicine precision (Lange et al. 2018; Leonard et al. 2018; Rello et al. 2018). Thus, the control of the various variables makes ASTs expensive both in healthcare centers and in biomedical research, which is why new theranostics approaches capable of selecting the appropriate therapy and measuring the response to treatment will be of high impact (Sahlgren et al. 2017; Syal et al. 2017). In this way, a review of the application of the concept of antimicrobial resistance that transcends phenotypic and genotypic adaptations to antibiotics by pathogens should be carried out in order to develop a new test model that allows timely diagnosis and adequate treatment of affected patients (Li and Webster 2018). Thus, the objective of this chapter is to analyze the fundamentals of antimicrobial susceptibility tests in order to project this important tool toward a field of research applied in theranostics.

1.2 Antimicrobial Susceptibility Testing from Anti-infectives Research Until Clinical Use

Perhaps the first use of antibiotic susceptibility testing was for the discovery and development of new drugs prior to reports of the resistance phenomenon that prompted the implementation of protocols with greater clinical prediction capacity (Gajdács et al. 2017; Landecker 2019). Thus, the first antimicrobial tests were aimed at identifying active compounds through in vivo infection models, where it was considered that the promising molecule could increase the survival rate of the test organism (Schumacher et al. 2018; Maugeri et al. 2019). In these first approaches, it was very possible to rule out promising active molecules due to the simple fact of not presenting a correct absorption and distribution in the infected tissues, so the success rate in the discovery of new anti-infective drugs was reduced and it was urgent to develop a model of primary screening to select antimicrobial agents before taking them to the animal model (O'Connell et al. 2013; Moffat et al. 2017; Kaczor et al. 2018). In this way, one of the first experiments of in vitro antimicrobial activity was carried out by Alexander Fleming in 1928 in order to demonstrate the ability of fungi of the *Penicillium* species to inhibit various types of bacteria in a petri dish (Sherpa et al. 2015; Arseculeratne and Arseculeratne 2017; Nicolaou and Rigol 2018). It was also Fleming himself in his speech to receive the Nobel Prize who warned of the possibility that microorganisms develop resistance when exposed to suboptimal doses of antibiotics, which finally happened during the golden era of antimicrobial chemotherapy (Yap 2013; Ventola 2015). Thus, given the emergence of anti-infective resistance, it was necessary to implement and standardize reproducible and precise methods capable of determining the susceptibility of microorganisms to medications as well as being a fundamental tool in the discovery and development of new medications (Castro-Pastrana et al. 2016; Centeno-Leija et al. 2016). In this way, the first approaches by an in vitro technique capable of predicting the clinical response required taking into consideration the distribution of drugs in the tissues, as well as the optimal inhibitory dose of the pathogens against the antimicrobials, as well as the amount of ideal microorganisms to be exposed to the antibiotic without altering the results was considered (Sy and Derendorf 2016; Stratton 2018). Finally, as a result of the initial correlations between the mentioned variables, the phenotypic methods of antibacterial susceptibility were implemented as a guideline model for antimicrobial treatment as well as epidemiological surveillance of resistance (Connors et al. 2013; Schechner et al. 2013).

1.3 Phenotypic Resistance Versus Genetic Resistance

Phenotypic antimicrobial resistance is due to the adaptive capacity of microorganisms against the aggression represented by antibiotics; in this order of ideas, phenotypic resistance determines the persistence capacity of microbial populations, but not the post-adaptation mutation (Fig. 1.1) (Corona and Martinez 2013; Hughes and Andersson 2017; Jahn et al. 2017). Thus, genotypic resistance arises when the adaptive process seeks the reducing metabolic pathways in order to optimize the obtaining of energy that is altered under the stress phenomenon that the antimicrobial treatment develops (Wales and Davies 2015; Yen and Papin 2017; Sandberg et al. 2019). Then, the phenotypic resistance is represented by the appearance of phenomena such as the decrease of the permeability of the microbial cell by thickening of the wall, as well as the formation of biofilms capable of decreasing the intracellular concentration of the antimicrobial and allowing persistence (Beceiro et al. 2013; Singh et al. 2017; Ciofu and Tolker-Nielsen 2019). Likewise, the expression of efflux pumps that expel antibiotics out of the cell facilitates the survival of microorganisms although this leads to a large expenditure of energy that requires that metabolic pathways be optimized and

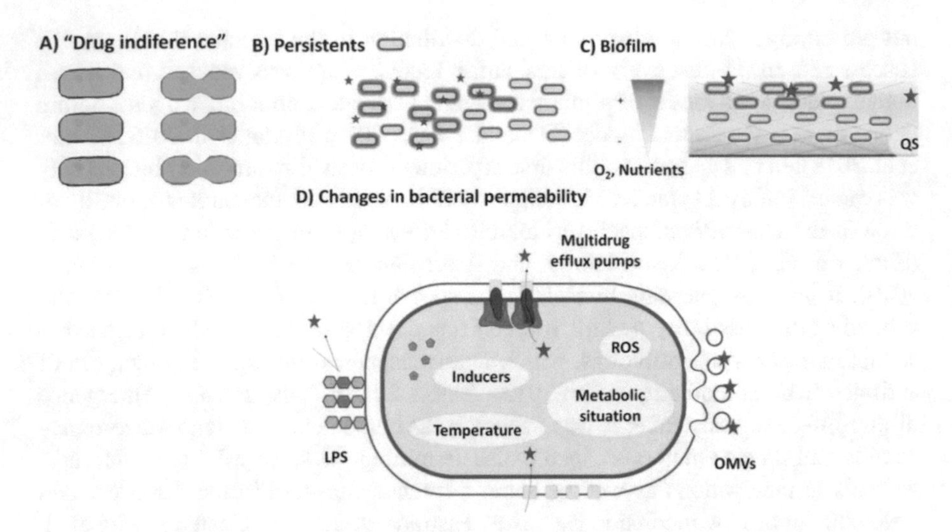

Fig. 1.1 Mechanisms of phenotypic resistance

the replication of genetic material is accelerated, at this point. This is the point where the phenotypic resistance acquires mutations to acquire a resistance that can be transmitted to the following generations, but this requires adaptation in fitness (Fair and Tor 2014; Culyba et al. 2015). Following this phenomenon, genotypic resistance appears when mutations in the genome replication that express the various therapeutic targets of antimicrobials are accumulated, causing variants that allow energy to be obtained from cells under stress, so continued use and improper anti-infectives allow mutations to persist and optimize microbial metabolism (Carlson-Banning and Zechiedrich 2015; Munita and Arias 2016; Schroeder et al. 2017). Currently, antimicrobial screening tests belong to the type of phenotypic test that detects the persistence of microorganisms against antibiotics; genotypic tests are more useful to determine the spread of resistance and distribution of multiresistant strains (Ellington et al. 2017; Kaur and Peterson 2018; Klümper et al. 2019). It is also important to take into consideration that antimicrobial tests evaluate susceptibility in an antimicrobial in vitro model, which does not happen in the human ecosystem where the microbiome modulates the different populations in contact with the host, so the phenotypic persistence of the microorganism remains a more reliable measure of the feasibility of a treatment (Gjini and Brito 2016; Foster et al. 2017; Milani et al. 2017).

1.4 Antimicrobial Susceptibility Testing and Virulence Factors

One of the aspects that antimicrobial susceptibility tests cannot assess is the influence of antibiotics on the virulence factors of microorganisms, which could determine the course of the clinical response of the disease (Fig. 1.2) (Doern and Brecher

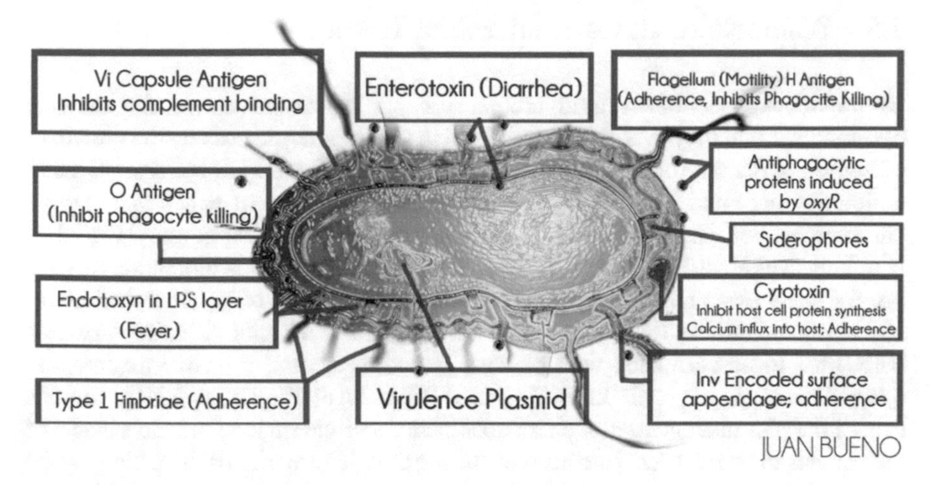

Fig. 1.2 Bacterial virulence factors

2011; Bengtsson-Palme et al. 2017; Burnham et al. 2017). Equally important is the implementation of a susceptibility technique capable of correlating antibiofilm activity with the response to antimicrobial therapy which has not been possible due to the high concentrations required by a biofilm to be inhibited in vitro by clinical practice antibiotics (Lebeaux et al. 2014; Macia et al. 2014; Baptista et al. 2018). In this way, the susceptibility tests only evaluate the activity on cells in suspension and not on microbial formations that may be the niche of resistant mutants, due to the exchange of genetic information that occurs inside the biofilms as well as their strength against environmental factors (Azeredo et al. 2017; Cattò and Cappitelli 2019; Liu et al. 2019). Thus, ASTs do not have the ability to determine the activity or inhibitory capacity of an antimicrobial molecule or formulation on the virulence factors of the pathogen, what diminishes its potential in the discovery and development of new antimicrobials, as well as in the control of the appearance of resistance (Rios et al. 2016; Wang et al. 2017; Kamaruzzaman et al. 2019). Similarly, ASTs do not predict the correct absorption and distribution of the drug in the patient; they simply indicate the concentration that inhibits the microorganism, and although it is possible to discriminate between susceptible and resistant organisms with this information, the possibility of developing resistance during treatment is not predicted (Smith and Kirby 2016; Buckheit and Lunsford 2017; Thaler et al. 2019). On the other hand, ASTs that detect the presence of resistance genes in infectious agents do not determine the biological and adaptive capacity of existing mutations; they also do not detect the ability of adhesion and biofilm formation of organisms that infect the patient (Römling and Balsalobre 2012; Bjarnsholt et al. 2013; Gebreyohannes et al. 2019). Due to the aforementioned, it is necessary to implement theranostics approaches and concepts of precision medicine in the antimicrobial susceptibility protocols that allow more information on the cure of the patient during the course of the infectious disease (Marciello et al. 2016; Knowles et al. 2017; Miller et al. 2019).

1.5 Biomarkers and Antimicrobial Testing

An important aspect of the antimicrobial susceptibility tests is concerning the type of biomarker that they evaluate, represented in the phenotypic tests by the growth or inhibition of the microorganism in vitro culture when exposed to concentrations of antibiotic drugs and in the genotypic tests by the presence of genes that express antimicrobial resistance mechanisms that configure the resistome (Fig. 1.3) (Frickmann et al. 2014; Choi et al. 2014; Khan et al. 2019). This biomarker of wide use for many years has disadvantages to predict the clinical response to antimicrobial treatment, where the minimum concentration that inhibits the microorganism (MIC) has to be correlated with the cure rate with the anti-infective treatment at optimal doses (Dijck et al. 2018; Veiga and Paiva 2018; Tsuji et al. 2019), and this causes the pharmacogenomic, pharmacokinetic, and pharmacodynamic aspects of the patients to be not taken into account during the performance of the antimicrobial tests and also when trying to discover new compounds that prevent selecting the promising molecules that will be in a formulation for clinical use (Calvo et al. 2016; Fermini et al. 2018; Guthrie and Kelly 2019). Thus, MICs as a biomarker of the activity and antimicrobial resistance must be rethought to a more comprehensive measure that correlates the environmental factors, the ecology of the microorganism, and the functional biology of the patient, as well as detecting the ability to select the most effective treatment either under a prospecting protocol or within healthcare units (Ferri et al. 2017; Kraemer et al. 2019). Likewise, MICs should be a biomarker capable of predicting the selection of resistant mutants by using suboptimal doses of antibiotics, as the result of phenotypic and genotypic resistance mechanisms within the microbial cell in response to stress induced by antimicrobial treatment (Cairns et al. 2018; Nguyen et al. 2018; Blondeau and Fitch 2019). In this way, the antimicrobial resistance (AMR) genes can have a great predictive value in

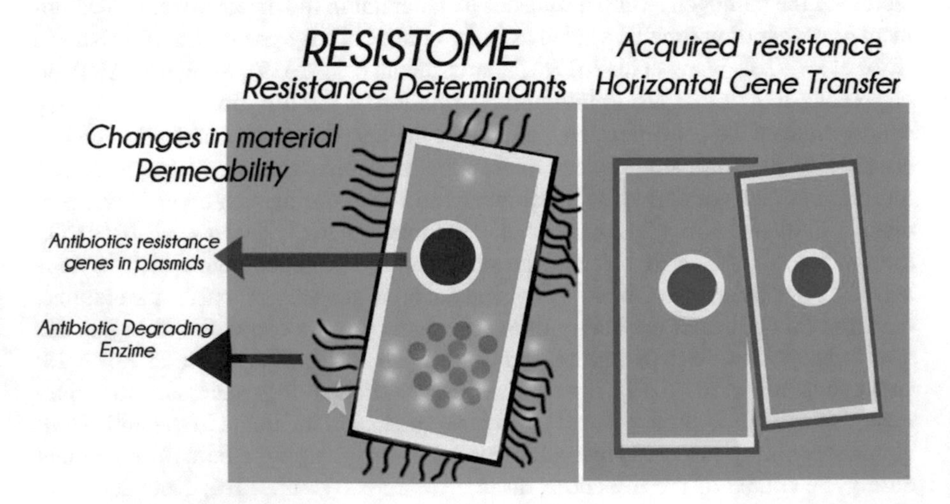

Fig. 1.3 Microbial resistome and resistance mechanisms

defining the treatment that will possibly fail but not in determining the successful therapy, given that biomarkers of antimicrobial persistence are required, as well as those of hormesis that allow to know what subinhibitory doses can cause increased microbial replication (Marston et al. 2016; Dougan et al. 2019; Hau et al. 2018). Thus, among the biomarkers of antimicrobial persistence that confer phenotypic resistance are various virulence factors such as adhesion factors, evasion of the immune response, the ability to form biofilms, and permeability to anti-infectives, necessitating the implementation of a comprehensive model of antimicrobial susceptibility (Bengoechea and Sa Pessoa 2018; Van Gerven et al. 2018; Su et al. 2018).

1.6 Integral Antimicrobial Susceptibility Testing

Until now ASTs are very effective in predicting the therapy that will fail, but not in predicting therapeutic success, which prevents a rational use of antibiotics and gives an important margin for the appearance of AMR (Arslan et al. 2017; Hawkey et al. 2018). In this order of ideas, a comprehensive AST platform is required that performs a multivariate analysis and allows the integration of MIC, pharmacokinetics, microbial persistence, and the ability to prevent the formation of resistant mutants (Zhao and Drlica 2001; Cohen et al. 2013; Davies and Wales 2019). But this approach must also be combined with a protocol that allows rapid, reproducible, and automatable results of both clinical samples and isolated microorganisms, in order for clinical doctors to make rational decisions in antibiotic management, epidemiologists to build maps of dissemination and risk of the disease, as well as researchers in pharmacology to increase the potency of high-throughput screening (Maurer et al. 2017; Mauri et al. 2017; Alamgir 2018). In this way, theranostics approaches that combine markers aimed at the specific target, together with antimicrobial treatment, can be the key to designing antimicrobial susceptibility models that can give information on resistance, virulence factors, and pharmacogenetics in proof (Ahmed et al. 2014a, b; Jeelani et al. 2014; Siest and Schallmeiner 2014). Likewise, the possibility of determining the type of infectious disease and selecting a treatment that does not induce resistance is one of the most promising applications of theranostics, which allows not only the use of antimicrobials of chemical origin but also of physics such as light (Mallidi et al. 2016; Patra et al. 2018; He et al. 2019).

1.7 Biosensors in Antimicrobial Evaluation

In this order of ideas, to make theranostics platform automatable, it is necessary to develop biosensors that can be directed to the infectious focus and from there determine the type of microorganism and the response to therapy in situ (Park et al. 2013; Ahmed et al. 2014a, b; Walper et al. 2018). Thus, the new antimicrobial susceptibility methods should contemplate the application of biosensors that can detect proteins, genetic material, and virulence factors of microbial origin, together with biomarkers of pharmacokinetics such as the presence of transport proteins and cytochromes (Bueno 2014; Metzger et al. 2017). In this sense, any diagnostic method of antibiotic treatment selection must be able to determine clinical endpoints that allow predicting the response to therapy, and extensive correlation studies are required to demonstrate the application of biosensors for the discovery of new drugs and hospital care (Mehrotra 2016; Weiss and Nowak-Sliwinska 2017; Rawson et al. 2019).

1.8 Conclusions

Antimicrobial susceptibility tests should evolve toward precision medicine as a fundamental requirement to avoid the emergence of antimicrobial resistance, which is one of the major threats in public health (Prestinaci et al. 2015; Lhermie et al. 2017). In this development, theranostics will play a fundamental role in finding an adequate clinical correlation that allows selecting the appropriate treatment for each infectious disease (Caliendo et al. 2013). Likewise, the search for predictive biomarkers for the design and development of new biosensors that can detect the presence of infection and the virulence mechanisms of microorganisms is necessary to be able to obtain high reproducibility and robustness devices (Vidic et al. 2017). Likewise, the prediction of the pharmacokinetics and pharmacodynamics of antimicrobial drugs is a field where precision personalized medicine will allow defining the parameters that will determine if the therapy will be successful, which is the great end and achievement of future susceptibility tests (Onufrak et al. 2016).

Acknowledgments The author thanks Sebastian Ritoré for his collaboration and invaluable support during the writing of this chapter, as well as the graphics contained in this book.

References

Ahmed, A., Rushworth, J. V., Hirst, N. A., & Millner, P. A. (2014a). Biosensors for whole-cell bacterial detection. *Clinical Microbiology Reviews, 27*(3), 631–646.

Ahmed, M. U., Saaem, I., Wu, P. C., & Brown, A. S. (2014b). Personalized diagnostics and biosensors: A review of the biology and technology needed for personalized medicine. *Critical Reviews in Biotechnology, 34*(2), 180–196.

Alamgir, A. N. M. (2018). Molecular pharmacognosy—A new borderline discipline between molecular biology and pharmacognosy. In *Therapeutic use of medicinal plants and their extracts* (Vol. 2, pp. 665–720). Cham: Springer.

Arseculeratne, S. N., & Arseculeratne, G. (2017). A re-appraisal of the conventional history of antibiosis and penicillin. *Mycoses, 60*(5), 343–347.

Arslan, N., Yılmaz, Ö., & Demiray-Gürbüz, E. (2017). Importance of antimicrobial susceptibility testing for the management of eradication in Helicobacter pylori infection. *World Journal of Gastroenterology, 23*(16), 2854.

Azeredo, J., Azevedo, N. F., Briandet, R., Cerca, N., Coenye, T., Costa, A. R., et al. (2017). Critical review on biofilm methods. *Critical Reviews in Microbiology, 43*(3), 313–351.

Balouiri, M., Sadiki, M., & Ibnsouda, S. K. (2016). Methods for in vitro evaluating antimicrobial activity: A review. *Journal of Pharmaceutical Analysis, 6*(2), 71–79.

Baptista, P. V., McCusker, M. P., Carvalho, A., Ferreira, D. A., Mohan, N. M., Martins, M., & Fernandes, A. R. (2018). Nano-strategies to fight multidrug resistant bacteria—"A Battle of the Titans". *Frontiers in Microbiology, 9.*

Beceiro, A., Tomás, M., & Bou, G. (2013). Antimicrobial resistance and virulence: A successful or deleterious association in the bacterial world? *Clinical Microbiology Reviews, 26*(2), 185–230.

Bengoechea, J. A., & Sa Pessoa, J. (2018). Klebsiella pneumoniae infection biology: Living to counteract host defences. *FEMS Microbiology Reviews, 43*(2), 123–144.

Bengtsson-Palme, J., Kristiansson, E., & Larsson, D. J. (2017). Environmental factors influencing the development and spread of antibiotic resistance. *FEMS Microbiology Reviews, 42*(1), fux053.

Bjarnsholt, T., Ciofu, O., Molin, S., Givskov, M., & Høiby, N. (2013). Applying insights from biofilm biology to drug development—can a new approach be developed? *Nature Reviews Drug Discovery, 12*(10), 791–808.

Blondeau, J. M., & Fitch, S. D. (2019). Mutant prevention and minimum inhibitory concentration drug values for enrofloxacin, ceftiofur, florfenicol, tilmicosin and tulathromycin tested against swine pathogens Actinobacillus pleuropneumoniae, Pasteurella multocida and Streptococcus suis. *PLoS One, 14*(1), e0210154.

Buckheit, R. W., & Lunsford, R. D. (2017). *In vitro* performance and analysis of combination anti-infective evaluations. In *Antimicrobial drug resistance* (pp. 1329–1345). Cham: Springer.

Bueno, J. (2014). Biosensors in antimicrobial drug discovery: Since biology until screening platforms. *Journal of Microbial & Biochemical Technology, S10.*

Burnham, C. A. D., Leeds, J., Nordmann, P., O'Grady, J., & Patel, J. (2017). Diagnosing antimicrobial resistance. *Nature Reviews Microbiology, 15*(11), 697.

Cairns, J., Ruokolainen, L., Hultman, J., Tamminen, M., Virta, M., & Hiltunen, T. (2018). Ecology determines how low antibiotic concentration impacts community composition and horizontal transfer of resistance genes. *Communications Biology, 1*(1), 35.

Caliendo, A. M., Gilbert, D. N., Ginocchio, C. C., Hanson, K. E., May, L., Quinn, T. C., et al. (2013). Better tests, better care: Improved diagnostics for infectious diseases. *Clinical Infectious Diseases, 57*(suppl_3), S139–S170.

Calvo, E., Walko, C., Dees, E. C., & Valenzuela, B. (2016). Pharmacogenomics, pharmacokinetics, and pharmacodynamics in the era of targeted therapies. *American Society of Clinical Oncology Educational Book, 36*, e175–e184.

Carlson-Banning, K. M., & Zechiedrich, L. (2015). Antibiotic classes and mechanisms of resistance. In *Encyclopedia of metagenomics: Environmental metagenomics* (pp. 20–31) Springer, New York, NY.

Castro-Pastrana, L. I., Serrano-Martínez, P., & Domínguez-Ramírez, L. (2016). Drug safety approaches in anti-infective drug discovery and development. In Atta-ur-Rahman (Ed.), *Frontiers in clinical drug research: Anti-infectives* (Vol. 2, pp. 95–136). Bentham Science. Sharjah, United Arab Emirates.

Cattò, C., & Cappitelli, F. (2019). Testing anti-biofilm polymeric surfaces: Where to start? *International Journal of Molecular Sciences, 20*(15), 3794.

Centeno-Leija, S., Guzmán-Trampe, S., Rodríguez-Peña, K., Bautista-Tovar, D., Espinosa, A., Trenado, M., & Sánchez, S. (2016). Different approaches for searching new microbial compounds with anti-infective activity. In *New weapons to control bacterial growth* (pp. 395–431). Cham: Springer.

Choi, J., Yoo, J., Lee, M., Kim, E. G., Lee, J. S., Lee, S., et al. (2014). A rapid antimicrobial susceptibility test based on single-cell morphological analysis. *Science Translational Medicine, 6*(267), 267ra174.

Ciofu, O., & Tolker-Nielsen, T. (2019). Tolerance and resistance of Pseudomonas aeruginosa biofilms to antimicrobial agents-How P. aeruginosa can escape antibiotics. *Frontiers in Microbiology, 10*, 913.

Cohen, N. R., Lobritz, M. A., & Collins, J. J. (2013). Microbial persistence and the road to drug resistance. *Cell Host and Microbe, 13*(6), 632–642.

Connors, K. P., Kuti, J. L., & Nicolau, D. P. (2013). Optimizing antibiotic pharmacodynamics for clinical practice. *Pharmaceutica Analytica Acta, 4*(3), 214.

Corona, F., & Martinez, J. L. (2013). Phenotypic resistance to antibiotics. *Antibiotics, 2*(2), 237–255.

Culyba, M. J., Mo, C. Y., & Kohli, R. M. (2015). Targets for combating the evolution of acquired antibiotic resistance. *Biochemistry, 54*(23), 3573–3582.

Davies, R., & Wales, A. (2019). Antimicrobial resistance on farms: A review including biosecurity and the potential role of disinfectants in resistance selection. *Comprehensive Reviews in Food Science and Food Safety, 18*(3), 753–774.

Dijck, P. V., Sjollema, J., Cammue, B. P., Lagrou, K., Berman, J., d'Enfert, C., et al. (2018). Methodologies for in vitro and in vivo evaluation of efficacy of antifungal and antibiofilm agents and surface coatings against fungal biofilms. *Microbial cell (Graz, Austria), 5*(7), 300.

Doern, G. V., & Brecher, S. M. (2011). The clinical predictive value (or lack thereof) of the results of in vitro antimicrobial susceptibility tests. *Journal of Clinical Microbiology, 49*(9 Supplement), S11–S14.

Dougan, G., Dowson, C., Overington, J., & Participants, N. G. A. D. S. (2019). Meeting the discovery challenge of drug-resistant infections: Progress and focusing resources. *Drug Discovery Today, 24*(2), 452–461.

Ellington, M. J., Ekelund, O., Aarestrup, F. M., Canton, R., Doumith, M., Giske, C., et al. (2017). The role of whole genome sequencing in antimicrobial susceptibility testing of bacteria: Report from the EUCAST Subcommittee. *Clinical Microbiology and Infection, 23*(1), 2–22.

Fair, R. J., & Tor, Y. (2014). Antibiotics and bacterial resistance in the 21st century. *Perspectives in Medicinal Chemistry, 6*, PMC-S14459.

Fermini, B., Coyne, S. T., & Coyne, K. P. (2018). Clinical trials in a dish: A perspective on the coming revolution in drug development. *SLAS DISCOVERY: Advancing Life Sciences RandD, 23*(8), 765–776.

Ferri, M., Ranucci, E., Romagnoli, P., & Giaccone, V. (2017). Antimicrobial resistance: A global emerging threat to public health systems. *Critical Reviews in Food Science and Nutrition, 57*(13), 2857–2876.

Foster, K. R., Schluter, J., Coyte, K. Z., & Rakoff-Nahoum, S. (2017). The evolution of the host microbiome as an ecosystem on a leash. *Nature, 548*(7665), 43–51.

Frickmann, H., Masanta, W. O., & Zautner, A. E. (2014). Emerging rapid resistance testing methods for clinical microbiology laboratories and their potential impact on patient management. *BioMed Research International, 2014*, 375681.

Gajdács, M., Spengler, G., & Urbán, E. (2017). Identification and antimicrobial susceptibility testing of anaerobic bacteria: Rubik's cube of clinical microbiology? *Antibiotics, 6*(4), 25.

Gebreyohannes, G., Nyerere, A., Bii, C., & Sbhatu, D. B. (2019). Challenges of intervention, treatment, and antibiotic resistance of biofilm-forming microorganisms. *Heliyon, 5*(8), e02192.

Gjini, E., & Brito, P. H. (2016). Integrating antimicrobial therapy with host immunity to fight drug-resistant infections: Classical vs. adaptive treatment. *PLoS Computational Biology, 12*(4), e1004857.

Guthrie, L., & Kelly, L. (2019). Bringing microbiome-drug interaction research into the clinic. *EBioMedicine, 44*, 708.

Hau, S. J., Haan, J. S., Davies, P. R., Frana, T., & Nicholson, T. L. (2018). Antimicrobial resistance distribution differs among methicillin resistant Staphylococcus aureus sequence type (ST) 5 isolates from health care and agricultural sources. *Frontiers in Microbiology, 9*, 2102.

Hawkey, P. M., Warren, R. E., Livermore, D. M., McNulty, C. A., Enoch, D. A., Otter, J. A., & Wilson, A. P. R. (2018). Treatment of infections caused by multidrug-resistant Gram-negative bacteria: report of the British Society for Antimicrobial Chemotherapy/healthcare Infection Society/british Infection Association Joint Working Party. *Journal of Antimicrobial Chemotherapy, 73*(suppl_3), iii2–iii78.

He, X., Xiong, L. H., Zhao, Z., Wang, Z., Luo, L., Lam, J. W. Y., et al. (2019). AIE-based theranostic systems for detection and killing of pathogens. *Theranostics, 9*(11), 3223.

Hughes, D., & Andersson, D. I. (2017). Environmental and genetic modulation of the phenotypic expression of antibiotic resistance. *FEMS Microbiology Reviews, 41*(3), 374–391.

Jahn, L. J., Munck, C., Ellabaan, M. M., & Sommer, M. O. (2017). Adaptive laboratory evolution of antibiotic resistance using different selection regimes lead to similar phenotypes and genotypes. *Frontiers in Microbiology, 8*, 816.

Jeelani, S., Reddy, R. J., Maheswaran, T., Asokan, G. S., Dany, A., & Anand, B. (2014). Theranostics: A treasured tailor for tomorrow. *Journal of Pharmacy and Bioallied Sciences, 6*(Suppl 1), S6.

Jenkins, S. G., & Schuetz, A. N. (2012). Current concepts in laboratory testing to guide antimicrobial therapy. In *Mayo Clinic proceedings* (Vol. 87, No. 3, pp. 290–308). Elsevier.

Kaczor, A. A., Medarametla, P., Bartuzi, D., Kondej, M., Matosiuk, D., & Poso, A. (2018). Molecular modelling approaches to antibacterial drug design and discovery. *Frontiers in Anti-Infective Drug Discovery, 7*(7), 153.

Kamaruzzaman, N. F., Tan, L. P., Hamdan, R. H., Choong, S. S., Wong, W. K., Gibson, A. J., et al. (2019). Antimicrobial polymers: The potential replacement of existing antibiotics? *International Journal of Molecular Sciences, 20*(11), 2747.

Kaur, P., & Peterson, E. (2018). Antibiotic resistance mechanisms in bacteria: Relationships between resistance determinants of antibiotic producers, environmental bacteria, and clinical pathogens. *Frontiers in Microbiology, 9*, 2928.

Khan, Z. A., Siddiqui, M. F., & Park, S. (2019). Current and emerging methods of antibiotic susceptibility testing. *Diagnostics (Basel), 9*(2), 49.

Kim, S., Masum, F., & Jeon, J. S. (2019). Recent developments of chip-based phenotypic antibiotic susceptibility testing. *BioChip Journal, 13*(1), 43–52.

Klümper, U., Recker, M., Zhang, L., Yin, X., Zhang, T., Buckling, A., & Gaze, W. H. (2019). Selection for antimicrobial resistance is reduced when embedded in a natural microbial community. *The ISME Journal 13*(12):2927–2937.

Knowles, L., Luth, W., & Bubela, T. (2017). Paving the road to personalized medicine: Recommendations on regulatory, intellectual property and reimbursement challenges. *Journal of Law and the Biosciences, 4*(3), 453–506.

Kraemer, S. A., Ramachandran, A., & Perron, G. G. (2019). Antibiotic pollution in the environment: From microbial ecology to public policy. *Microorganisms, 7*(6), 180.

Kurhekar, J., Tupas, G. D., & Otero, M. C. B. (2019). In-vitro assays for antimicrobial assessment. In *Phytochemistry: An in-silico and in-vitro update* (pp. 279–298). Singapore: Springer.

Landecker, H. (2019). Antimicrobials before antibiotics: War, peace, and disinfectants. *Palgrave Communications, 5*(1), 45.

Lange, C., Alghamdi, W. A., Al-Shaer, M. H., Brighenti, S., Diacon, A. H., DiNardo, A. R., et al. (2018). Perspectives for personalized therapy for patients with multidrug-resistant tuberculosis. *Journal of Internal Medicine, 284*(2), 163–188.

Lebeaux, D., Ghigo, J. M., & Beloin, C. (2014). Biofilm-related infections: Bridging the gap between clinical management and fundamental aspects of recalcitrance toward antibiotics. *Microbiology and Molecular Biology Reviews, 78*(3), 510–543.

Leonard, H., Colodner, R., Halachmi, S., & Segal, E. (2018). Recent advances in the race to design a rapid diagnostic test for antimicrobial resistance. *ACS Sensors, 3*(11), 2202–2217.

Lhermie, G., Gröhn, Y. T., & Raboisson, D. (2017). Addressing antimicrobial resistance: An overview of priority actions to prevent suboptimal antimicrobial use in food-animal production. *Frontiers in Microbiology, 7*, 2114.

Li, B., & Webster, T. J. (2018). Bacteria antibiotic resistance: New challenges and opportunities for implant-associated orthopedic infections. *Journal of Orthopaedic Research®, 36*(1), 22–32.

Li, J., Xie, S., Ahmed, S., Wang, F., Gu, Y., Zhang, C., et al. (2017). Antimicrobial activity and resistance: Influencing factors. *Frontiers in Pharmacology, 8*, 364.

Liu, Y., Shi, L., Su, L., van der Mei, H. C., Jutte, P. C., Ren, Y., & Busscher, H. J. (2019). Nanotechnology-based antimicrobials and delivery systems for biofilm-infection control. *Chemical Society Reviews, 48*(2), 428–446.

Macia, M. D., Rojo-Molinero, E., & Oliver, A. (2014). Antimicrobial susceptibility testing in biofilm-growing bacteria. *Clinical Microbiology and Infection, 20*(10), 981–990.

Mallidi, S., Anbil, S., Bulin, A. L., Obaid, G., Ichikawa, M., & Hasan, T. (2016). Beyond the barriers of light penetration: Strategies, perspectives and possibilities for photodynamic therapy. *Theranostics, 6*(13), 2458.

Marciello, M., Pellico, J., Fernandez-Barahona, I., Herranz, F., Ruiz-Cabello, J., & Filice, M. (2016). Recent advances in the preparation and application of multifunctional iron oxide and liposome-based nanosystems for multimodal diagnosis and therapy. *Interface Focus, 6*(6), 20160055.

Marston, H. D., Dixon, D. M., Knisely, J. M., Palmore, T. N., & Fauci, A. S. (2016). Antimicrobial resistance. *JAMA, 316*(11), 1193–1204.

Maugeri, G., Lychko, I., Sobral, R., & Roque, A. C. (2019). Identification and antibiotic-susceptibility profiling of infectious bacterial agents: A review of current and future trends. *Biotechnology Journal, 14*(1), 1700750.

Maurer, F. P., Christner, M., Hentschke, M., & Rohde, H. (2017). Advances in rapid identification and susceptibility testing of bacteria in the clinical microbiology laboratory: Implications for patient care and antimicrobial stewardship programs. *Infectious Disease Reports, 9*(1), 6839.

Mauri, C., Principe, L., Bracco, S., Meroni, E., Corbo, N., Pini, B., & Luzzaro, F. (2017). Identification by mass spectrometry and automated susceptibility testing from positive bottles: A simple, rapid, and standardized approach to reduce the turnaround time in the management of blood cultures. *BMC Infectious Diseases, 17*(1), 749.

Mehrotra, P. (2016). Biosensors and their applications–a review. *Journal of Oral Biology and Craniofacial Research, 6*(2), 153–159.

Menden, M. P., Wang, D., Mason, M. J., Szalai, B., Bulusu, K. C., Guan, Y., et al. (2019). Community assessment to advance computational prediction of cancer drug combinations in a pharmacogenomic screen. *Nature Communications, 10*(1), 2674.

Metzger, S. W., Howson, D. C., Goldberg, D. A., & Buttry, D. A. (2017). U.S. Patent No. 9,657,327. Washington, D.C.: U.S. Patent and Trademark Office.

Milani, C., Duranti, S., Bottacini, F., Casey, E., Turroni, F., Mahony, J., et al. (2017). The first microbial colonizers of the human gut: Composition, activities, and health implications of the infant gut microbiota. *Microbiology and Molecular Biology Reviews, 81*(4), e00036–e00017.

Miller, M. B., Atrzadeh, F., Burnham, C. A. D., Cavalieri, S., Dunn, J., Jones, S., et al. (2019). Clinical utility of advanced microbiology testing tools. *Journal of Clinical Microbiology, 57*(9), e00495–e00419.

Moffat, J. G., Vincent, F., Lee, J. A., Eder, J., & Prunotto, M. (2017). Opportunities and challenges in phenotypic drug discovery: An industry perspective. *Nature Reviews Drug Discovery, 16*(8), 531.

Munita, J. M., & Arias, C. A. (2016). Mechanisms of antibiotic resistance. *Microbiology Spectrum, 4*(2), 1–36.

Nguyen, M., Brettin, T., Long, S. W., Musser, J. M., Olsen, R. J., Olson, R., et al. (2018). Developing an in silico minimum inhibitory concentration panel test for Klebsiella pneumoniae. *Scientific Reports, 8*(1), 421.

Nicolaou, K. C., & Rigol, S. (2018). A brief history of antibiotics and select advances in their synthesis. *The Journal of Antibiotics, 71*(2), 153.

O'Connell, K. M., Hodgkinson, J. T., Sore, H. F., Welch, M., Salmond, G. P., & Spring, D. R. (2013). Combating multidrug-resistant bacteria: Current strategies for the discovery of novel antibacterials. *Angewandte Chemie International Edition, 52*(41), 10706–10733.

Onufrak, N. J., Forrest, A., & Gonzalez, D. (2016). Pharmacokinetic and pharmacodynamic principles of anti-infective dosing. *Clinical Therapeutics, 38*(9), 1930–1947.

Park, M., Tsai, S. L., & Chen, W. (2013). Microbial biosensors: Engineered microorganisms as the sensing machinery. *Sensors, 13*(5), 5777–5795.

Patra, J. K., Das, G., Fraceto, L. F., Campos, E. V. R., del Pilar Rodriguez-Torres, M., Acosta-Torres, L. S., et al. (2018). Nano based drug delivery systems: Recent developments and future prospects. *Journal of Nanobiotechnology, 16*(1), 71.

Prestinaci, F., Pezzotti, P., & Pantosti, A. (2015). Antimicrobial resistance: A global multifaceted phenomenon. *Pathogens and Global Health, 109*(7), 309–318.

Rawson, T. M., Gowers, S. A., Freeman, D. M., Wilson, R. C., Sharma, S., Gilchrist, M., et al. (2019). Microneedle biosensors for real-time, minimally invasive drug monitoring of phenoxymethylpenicillin: A first-in-human evaluation in healthy volunteers. *The Lancet Digital Health, 1*(7), e335–e343.

Rello, J., Van Engelen, T. S. R., Alp, E., Calandra, T., Cattoir, V., Kern, W. V., et al. (2018). Towards precision medicine in sepsis: A position paper from the European Society of Clinical Microbiology and Infectious Diseases. *Clinical Microbiology and Infection, 24*(12), 1264–1272.

Rios, A. C., Moutinho, C. G., Pinto, F. C., Del Fiol, F. S., Jozala, A., Chaud, M. V., et al. (2016). Alternatives to overcoming bacterial resistances: State-of-the-art. *Microbiological Research, 191*, 51–80.

Römling, U., & Balsalobre, C. (2012). Biofilm infections, their resilience to therapy and innovative treatment strategies. *Journal of Internal Medicine, 272*(6), 541–561.

Sahlgren, C., Meinander, A., Zhang, H., Cheng, F., Preis, M., Xu, C., et al. (2017). Tailored approaches in drug development and diagnostics: From molecular design to biological model systems. *Advanced Healthcare Materials, 6*(21), 1700258.

Sandberg, T. E., Salazar, M. J., Weng, L. L., Palsson, B. O., & Feist, A. M. (2019). The emergence of adaptive laboratory evolution as an efficient tool for biological discovery and industrial biotechnology. *Metabolic Engineering, 56*, 1.

Schechner, V., Temkin, E., Harbarth, S., Carmeli, Y., & Schwaber, M. J. (2013). Epidemiological interpretation of studies examining the effect of antibiotic usage on resistance. *Clinical Microbiology Reviews, 26*(2), 289–307.

Schroeder, M., Brooks, B. D., & Brooks, A. E. (2017). The complex relationship between virulence and antibiotic resistance. *Genes, 8*(1), 39.

Schumacher, A., Vranken, T., Malhotra, A., Arts, J. J. C., & Habibovic, P. (2018). In vitro antimicrobial susceptibility testing methods: Agar dilution to 3D tissue-engineered models. *European Journal of Clinical Microbiology and Infectious Diseases, 37*(2), 187–208.

Sherpa, R. T., Reese, C. J., & Aliabadi, H. M. (2015). Application of iChip to grow "unculti-vable" microorganisms and its impact on antibiotic discovery. *Journal of Pharmacy and Pharmaceutical Sciences, 18*(3), 303–315.

Siest, G., & Schallmeiner, E. (2014). Pharmacogenomics and Theranostics in Practice: A summary of the Euromedlab-ESPT (The European Society of Pharmacogenomics and Theranostics) satellite symposium, May 2013. *EJIFCC, 24*(3), 85.

Singh, S., Singh, S. K., Chowdhury, I., & Singh, R. (2017). Understanding the mechanism of bacterial biofilms resistance to antimicrobial agents. *The Open Microbiology Journal, 11*, 53.

Smith, K. P., & Kirby, J. E. (2016). Validation of a high-throughput screening assay for identification of adjunctive and directly acting antimicrobials targeting carbapenem-resistant Enterobacteriaceae. *Assay and Drug Development Technologies, 14*(3), 194–206.

Stratton, C. W. (2018). Advanced phenotypic antimicrobial susceptibility testing methods. In *Advanced techniques in diagnostic microbiology* (pp. 69–98). Cham: Springer.

Su, Y. C., Jalalvand, F., Thegerström, J., & Riesbeck, K. (2018). The interplay between immune response and bacterial infection in COPD: Focus upon non-typeable Haemophilus influenzae. *Frontiers in Immunology, 9*, 2530.

Sy, S. K., & Derendorf, H. (2016). Pharmacokinetics I: PK-PD approach, the case of antibiotic drug development. In *Clinical pharmacology: Current topics and case studies* (pp. 185–217). Cham: Springer.

Syal, K., Mo, M., Yu, H., Iriya, R., Jing, W., Guodong, S., et al. (2017). Current and emerging techniques for antibiotic susceptibility tests. *Theranostics, 7*(7), 1795.

Tannert, A., Grohs, R., Popp, J., & Neugebauer, U. (2019). Phenotypic antibiotic susceptibility testing of pathogenic bacteria using photonic readout methods: Recent achievements and impact. *Applied Microbiology and Biotechnology, 103*(2), 549–566.

Thaler, D. S., Head, M. G., & Horsley, A. (2019). Precision public health to inhibit the contagion of disease and move toward a future in which microbes spread health. *BMC Infectious Diseases, 19*(1), 120.

Tsuji, B. T., Pogue, J. M., Zavascki, A. P., Paul, M., Daikos, G. L., Forrest, A., et al. (2019). International consensus guidelines for the optimal use of the polymyxins: Endorsed by the American College of Clinical Pharmacy (ACCP), European Society of Clinical Microbiology and Infectious Diseases (ESCMID), Infectious Diseases Society of America (IDSA), International Society for Anti-infective Pharmacology (ISAP), Society of Critical Care Medicine (SCCM), and Society of Infectious Diseases Pharmacists (SIDP). *Pharmacotherapy: The Journal of Human Pharmacology and Drug Therapy, 39*(1), 10–39.

Van Gerven, N., Van der Verren, S. E., Reiter, D. M., & Remaut, H. (2018). The role of functional amyloids in bacterial virulence. *Journal of Molecular Biology, 430*(20), 3657–3684.

Veiga, R. P., & Paiva, J. A. (2018). Pharmacokinetics–pharmacodynamics issues relevant for the clinical use of beta-lactam antibiotics in critically ill patients. *Critical Care, 22*(1), 233.

Ventola, C. L. (2015). The antibiotic resistance crisis: Part 1: Causes and threats. *Pharmacy and Therapeutics, 40*(4), 277.

Vidic, J., Manzano, M., Chang, C. M., & Jaffrezic-Renault, N. (2017). Advanced biosensors for detection of pathogens related to livestock and poultry. *Veterinary Research, 48*(1), 11.

Volz, C., Ramoni, J., Beisken, S., Galata, V., Keller, A., Plum, A., et al. (2019). Clinical resistome screening of 1,110 Escherichia coli isolates efficiently recovers diagnostically relevant antibiotic resistance biomarkers and potential novel resistance mechanisms. *Frontiers in Microbiology, 10*, 1671.

Wales, A. D., & Davies, R. H. (2015). Co-selection of resistance to antibiotics, biocides and heavy metals, and its relevance to foodborne pathogens. *Antibiotics, 4*(4), 567–604.

Walper, S. A., Lasarte Aragonés, G., Sapsford, K. E., Brown, C. W., III, Rowland, C. E., Breger, J. C., & Medintz, I. L. (2018). Detecting biothreat agents: From current diagnostics to developing sensor technologies. *ACS Sensors, 3*(10), 1894–2024.

Wang, L., Hu, C., & Shao, L. (2017). The antimicrobial activity of nanoparticles: Present situation and prospects for the future. *International Journal of Nanomedicine, 12*, 1227.

Weiss, A., & Nowak-Sliwinska, P. (2017). Current trends in multidrug optimization: An alley of future successful treatment of complex disorders. *SLAS TECHNOLOGY: Translating Life Sciences Innovation, 22*(3), 254–275.

Yap, M. N. F. (2013). The double life of antibiotics. *Missouri Medicine, 110*(4), 320.

Yen, P., & Papin, J. A. (2017). History of antibiotic adaptation influences microbial evolutionary dynamics during subsequent treatment. *PLoS Biology, 15*(8), e2001586.

Zhao, X., & Drlica, K. (2001). Restricting the selection of antibiotic-resistant mutants: A general strategy derived from fluoroquinolone studies. *Clinical Infectious Diseases, 33*(Supplement_3), S147–S156.

Chapter 2
Antimicrobial Activity of Nanomaterials: From Selection to Application

When it comes to atoms, language can be used only as in poetry. The poet, too, is not nearly so concerned with describing facts as with creating images
—Niels Bohr (1885–1962)

I believe that the same process of moulding of plastic materials into a configuration complementary to that of another molecule, which serves as a template, is responsible for all biological specificity. I believe that the genes serve as the templates on which are moulded the enzymes that are responsible for the chemical characters of the organisms, and that they also serve as templates for the production of replicas of themselves. The detailed mechanism by means of which a gene or a virus molecule produces replicas of itself is not yet known. In general the use of a gene or virus as a template would lead to the formation of a molecule not with identical structure but with complementary structure. It might happen, of course, that a molecule could be at the same time identical with and complementary to the template on which it is moulded
—Linus Pauling (1901–1994)

Abstract The correct application of nanotechnology for the treatment and control of antimicrobial resistance requires the implementation of translational science protocols capable of selecting the most active products with the lowest toxicity and that can be applied in the different areas where they may be needed. In this order of ideas, the selection of a reproducible antimicrobial platform that can be scalable, robust, and automatable becomes a necessity to be solved by nanotechnology researchers. Thus, the design and development of nanomaterials should be together with the implementation of evaluation and toxicity protocols in order to determine the promising nanoproducts to be used in the different innovations. For this reason, the objective of this chapter is to analyze the elements to be considered in the selection, implementation, and standardization of the necessary platforms to translate

J. Bueno, *Preclinical Evaluation of Antimicrobial Nanodrugs*, Nanotechnology in the Life Sciences, https://doi.org/10.1007/978-3-030-43855-5_2

nanotechnology to medicine, industry, and agriculture, as well as the protocols implemented to obtain an adequate therapeutic index of the products obtained in order to increase their efficacy and safety.

2.1 Introduction

In this section, it is very important to comment and define: What is antimicrobial activity? What causes a compound or material to be considered antimicrobial? The response from the point of view of susceptibility tests can be defined as activity inhibiting the growth of a microorganism, as well as the death of a microbial population below three logarithms of the initial population when exposed to certain concentrations (Armas et al. 2019; Hanson et al. 2019; Zoffmann et al. 2019). In this order of ideas, the inhibition of virulence factors such as adhesion, toxin production, immunosuppression, and biofilm formation can also be considered antimicrobial activity, because although they do not determine the destruction of the microorganism, they prevent the invasion and production of the infectious disease (Kostakioti et al. 2013; Chiu et al. 2017; Sabaté Brescó et al. 2017; Yu et al. 2018). Also within the antimicrobial activity should be considered the biocidal activity that measures the elimination of microorganisms in time ranges between 30 seconds and 20 minutes under the exposure of microbicidal agents; this aspect is essential to prevent the spread of infectious disease using antiseptics and disinfectants (Adlhart et al. 2018; Kramer et al. 2018; Deshmukh et al. 2019).Thus, the study and selection of nanomaterials for their antimicrobial activity require first assessing the use and application of nanotechnology in infectious diseases where it can be fundamental in the development of:

- Drugs carriers (Fig. 2.1)
- Devices with bioactivity
- Surfaces and equipment for protection against biothreats (Sampath Kumar and Madhumathi 2014; Wang et al. 2017; Burdușel et al. 2018)

In this order of ideas, the selection of the appropriate material for the treatment and control of infectious diseases should revolve around activity, toxicity, and application (Binas et al. 2017; Han and Ceilley 2017; Garcia et al. 2019). Likewise, in the application of the nanomaterial, it should be determined whether environmental toxicity is a factor to be considered to increase the safety of use as an anti-infective agent (Wolfram et al. 2015; Patra et al. 2018; Akter et al. 2018). For the above reasons, the objective of this chapter is to evaluate the criteria for the selection of safe and effective antimicrobial nanomaterials based on their application in medicine, agriculture, or industry (Álvarez-Paino et al. 2017; Jeevanandam et al. 2018; Adisa et al. 2019).

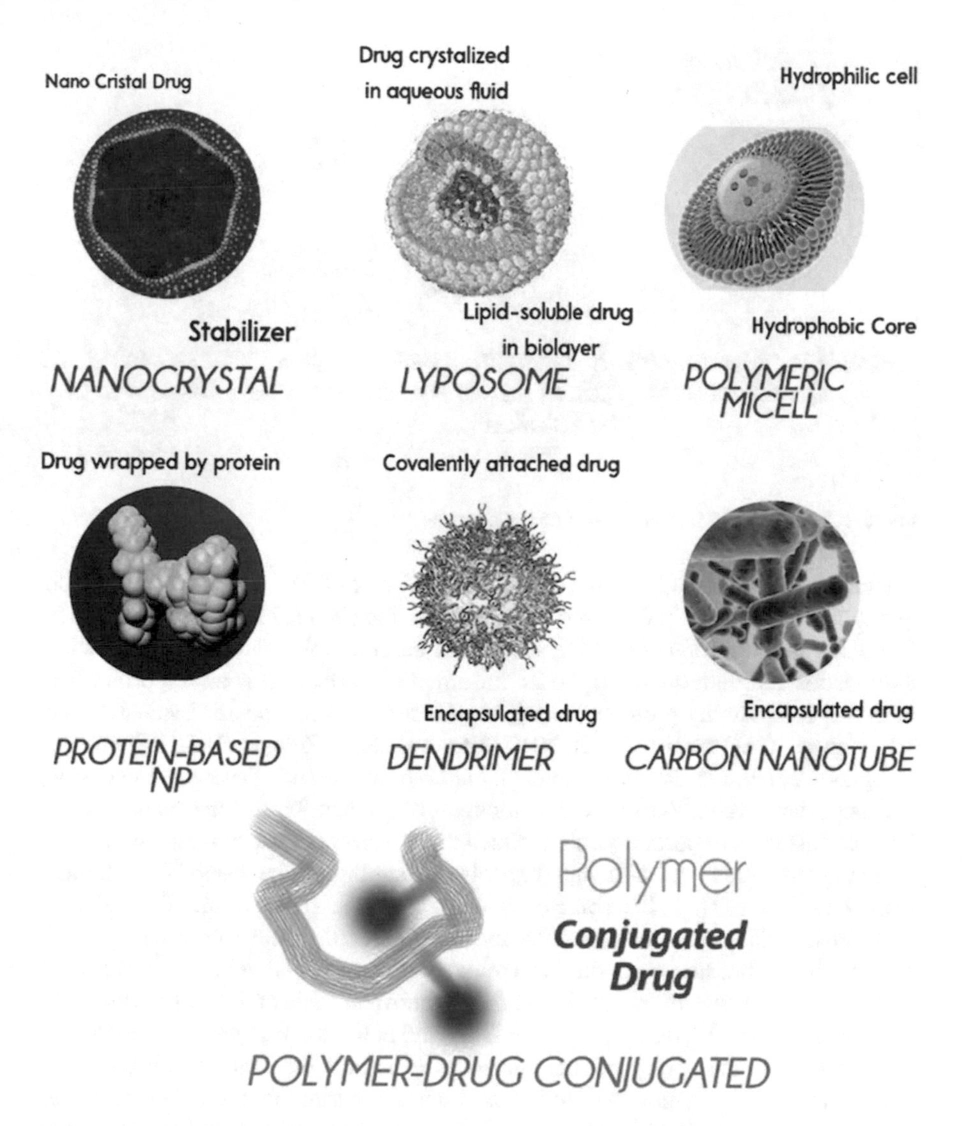

Fig. 2.1 Nanoantibiotics delivery systems

2.2 Antimicrobial Activity of Nanomaterials

The antimicrobial activity of nanomaterials can be divided depending on their application in anti-infective agents, antimicrobial adjuvants, and antimicrobial devices, as well as surfaces and biosafety materials (Fig. 2.2) (Raghunath and Perumal 2017; Han et al. 2018; Pezzi et al. 2019). In this order of ideas, the nanomaterial can be antimicrobial if it is used as the main microbicidal agent or as an adjuvant if it is used to favor the activity of a previously discovered antibiotic molecule; also the design and development of biocidal devices allows the use of the nanomaterial to

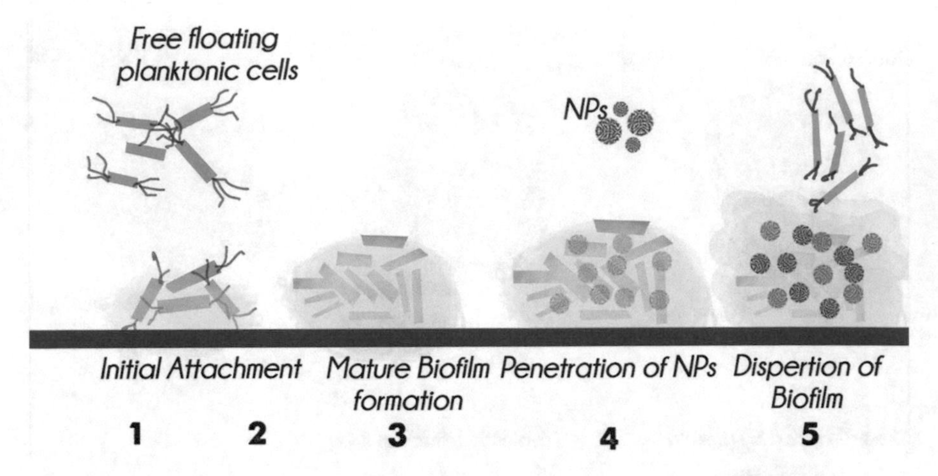

Fig. 2.2 Antimicrobial mechanisms of nanomaterials

perform invasive procedures in patients without risk of contamination and subsequent infection (de Melo Carrasco et al. 2015; Hemeg 2017; Makowski et al. 2019). Similarly, the development of functionalized antimicrobial surfaces with biocidal compounds and with the ability to inhibit microbial adhesion is an important field of action in obtaining protective equipment and requires a specific evaluation and selection system (Francolini et al. 2017; Koo et al. 2017; Zeng et al. 2018; Cattò and Cappitelli 2019). On the other hand, the antibiofilm activity becomes a necessary and important evaluation in the development of nanoantibiotics and nanobiocides because this is an important virulence factor that deserves to design molecules with greater penetration capacity and capable of inhibiting the established biofilm (Fig. 2.3) (Roy et al. 2018; Von Borowski et al. 2018; Torres et al. 2018; Sharma et al. 2019), although the evaluation models for biofilms have problems in their reproducibility because a standard microbial mass is not obtained for all procedures and there are always differences in the results between static and flow biofilm methods (Bueno 2014; Azeredo et al. 2017; Boudarel et al. 2018; Dijck et al. 2018). On the other hand, the ability of nanomaterials to inhibit virulence factors, such as adhesion factors, enzymes, and toxins, is another important factor to be determined because it prevents microbial invasion of tissues, so it should be classified as an anti-infective activity; thus, there is no destruction of pathogenic microorganisms (Mühlen and Dersch 2015; Munguia and Nizet 2017; Masters et al. 2019; Ruffin and Brochiero 2019).

2.3 Antimicrobial Models for Nanomaterials Evaluation

In this order of ideas, it is possible to take two routes for the evaluation of the antimicrobial activity of nanomaterials: one is that of antimicrobial susceptibility to obtain a minimum inhibitory concentration (MIC) and the other is the evaluation of

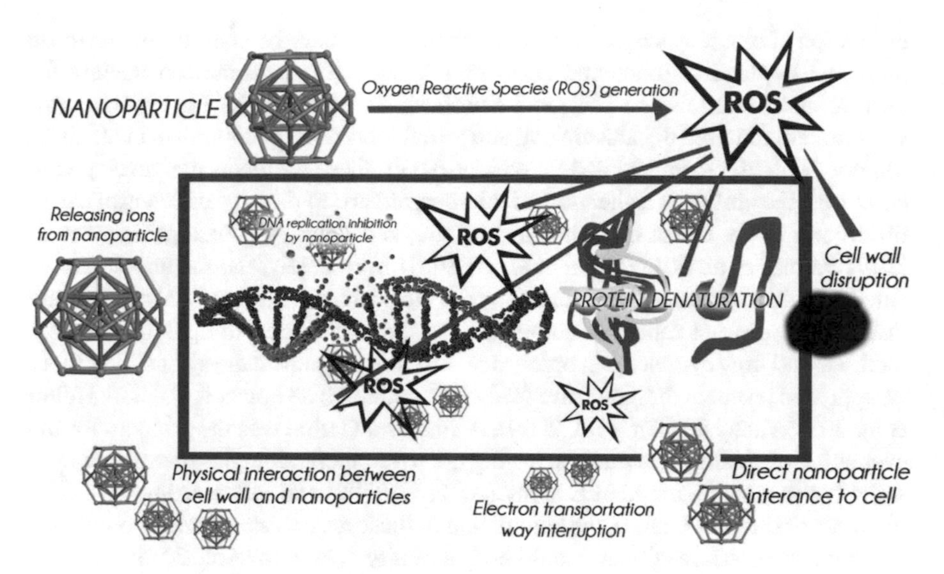

Fig. 2.3 Antibiofilm activity of nanomaterials

the biocidal activity to determine sterilizing, disinfectant, and antiseptic actions (Sirelkhatim et al. 2015; Yang et al. 2018). Thus, the protocols of the Clinical Laboratory Standards Institute (CLSI) and those of the European Committee on Antimicrobial Susceptibility Testing (EUCAST) for the evaluation of medicines and pure compounds have been implemented in the first way (Leonard et al. 2018; O'Halloran et al. 2018; Rancoita et al. 2018); both CLSI and EUCAST contemplate evaluation for bacteria, mycobacteria, and fungi using methods ranging from broth dilution to disk diffusion, and only CLSI contemplates a protocol for the evaluation of antiviral activity (Balouiri et al. 2016). Unfortunately, these methods have problems for the evaluation of new molecules and formulations due to their low automation in the case of broth dilution, such as the low diffusion of nonpolar compounds in the agar as in the case of disk diffusion, which affects the evaluation of antimicrobial activity (Egorova et al. 2017; Stratton 2018; Eloff 2019). It has also tried to automate these techniques using fluorescent dyes that determine cell viability such as resazurin, SYTO9, and 4′,6-diamino-2-phenylindole (DAPI) but that tend to react with metal nanoparticles, which alters the results of in vitro susceptibility (Panic et al. 2015; Magana et al. 2018). Thus, by virtue of the above, attempts have been made to implement more sensitive and specific methods such as flow cytometry and real-time PCR, in order to more accurately quantify the microbial mass after it has been exposed, but some materials such as nanotubes and graphene can inhibit the reactions necessary to obtain a correct reading (Emerson et al. 2017; Cossarizza et al. 2017; Kralik and Ricchi 2017; Cardano et al. 2018). So the major evaluation of nanomaterials antimicrobial activity using the CLSI and EUCAST protocols has been in the evaluation of nanotransporters of medications such as liposomes, micelles, and polymers (Álvarez-Paino et al. 2017; Sedghizadeh et al. 2017; Arévalo et al. 2019; Martin-Serrano et al. 2019). On the other hand, for the evaluation of biocidal activity, various methodologies have been used, such as those proposed by the Environmental Protection Agency (EPA), American Society for Testing and Materials (ASTM), and European Standards (EN) for evaluation of bactericidal, fungicidal, parasiticidal, and virucidal activity (Bloomfield et al. 2017; Malone et al. 2017; Soltani and Pouypouy 2019); these protocols are developed to determine the ability of a chemical or microorganisms to destroy and control harmful organisms for use in medicine, agriculture, and industry (Chattopadhyay et al. 2017; Parmar et al. 2017; Allah et al. 2019). Unfortunately, the methodologies to determine the biocidal activity continue with the problem of their visual reading and their techniques not robust or automated, so they have tried to apply fluorescent methods and flow cytometry in order to reduce these disadvantages but still without an approved protocol for global use (Tidwell et al. 2015; de Souza et al. 2016; Helmi et al. 2018; Vanhauteghem et al. 2019). A similar situation was the protocol for the evaluation of biofilms that despite having approved methodologies to evaluate both static biofilms using the MBEC assay (ASTM E2799) and inflow using CDC biofilm reactor (ASTM E2562), the automation of these tests to make them more reproducible and robust is still far from being a reality (Gazzola et al. 2015; Kerstens et al. 2015; Larimer et al. 2016; Bogachev et al. 2018). Thus, in this way, it is necessary to evaluate the different variables to be taken into account for the correct implementation of protocols for the evaluation of antimicrobial activity of nanomaterials (Chouhan et al. 2017; Khan et al. 2019; Kadiyala et al. 2018; Panpaliya et al. 2019).

2.4 Implementation of Antimicrobial Methods for Nanomaterials Evaluation

The first aspect to consider in the implementation of methods to evaluate the antimicrobial activity of nanomaterials is the physicochemical characteristics of the substances; thus, it is important to have information on solubility, chemical composition (metallic or not), polarity, and stability at different temperatures and conditions, as well as of the possible toxicity of the compound to be tested (Vimbela et al. 2017; Pavoni et al. 2019; Taghipour et al. 2019; Yusof et al. 2019). With the aforementioned characteristics, it is already possible to select the most suitable antimicrobial method (broth dilution or agar diffusion) and also the most appropriate reading technique (spectrophotometry, fluorescence, luminescence, or real-time PCR). It is important to note that the selected technique should always be reproducible, and then evaluate if it is possible to automate it and make it robust (Erdem et al. 2014; Horká et al. 2016; Wang and Salazar 2016; Luo et al. 2017). Another aspect to be considered is the application of the nanomaterial, the search for nanodrugs is very different than the design and development of antimicrobial covers for medical devices; as well as obtaining biocidal products for use in hospital, industrial, and agricultural environments (Singh et al. 2014; Burdușel et al. 2018; Thiruvengadam et al. 2018; Yang et al. 2019). The selection of the application of the nanomaterial acquires importance when standardizing the procedures of antimicrobial activity and in choosing the appropriate comparison controls to determine if the action is promising; using antibiotics or medications to compare pesticides is not appropriate, and the most widely tested control should be chosen always to have a greater chance of success (Jamil and Imran 2018; Kamaruzzaman et al. 2019; Ristić et al. 2019; Ruddaraju et al. 2020). On the other hand, the reading technique used to determine antimicrobial activity must be chosen for its sensitivity, specificity, and reproducibility, in order to avoid false-positive results, so it is sought to automate it to a greater extent; although spectrofluorimetric methods have been the most used, the metal nanoparticles alter the results, and this should be properly controlled (Azam et al. 2012; Vega-Jiménez et al. 2019). But it is important to remember that all activity selection criteria must be correlated with toxicity in order to determine that the product evaluated is active and safe; in that order of ideas with nanomaterials, it should also have special conditions (Rai et al. 2016; Drasler et al. 2017; Soeteman-Hernandez et al. 2019).

2.5 Selection Criteria on the Basis of Biodegradability and Toxicity

Among the fundamental aspects to take into account when determining the promising activity of any nanomaterial are its toxicity on human cells, as well as on animals and plants; it is also important to determine environmental toxicity in order to implement green nanomaterial synthesis protocols (Rana and Kalaichelvan 2013; Fadeel et al. 2018; Horky et al. 2018; Botha et al. 2019). Equally important in the field of nanotechnology is to include parameters that determine the primary skin as well as the lung toxicity, due to the degrees of exposure that these two organs will have in the future (Yildirimer et al. 2011; Kumar et al. 2014; Riediker et al. 2019). For this reason, it is necessary to implement the protocols recommended by the Organisation for Economic Co-operation and Development (OECD) in the laboratories for the evaluation of antimicrobial activity of nanomaterials, which include pulmonary bioassays, dermal irritation tests (OECD 404), acute oral toxicity test (OECD 425), and eye irritation test (OECD 405), as well as genotoxicity tests (OECD 471 and 473) and the battery of aquatic tests (OECD 203, 202 and 201) (Petersen et al. 2015; Dearfield et al. 2017; Savage et al. 2019; Siegrist et al. 2019). On the other hand, for protocols that determine the degradation, biodegradation, and transformation of nanomaterials, the European Chemicals Agency have standardized the methods set forth in document ECHA-17-G-15-EN to establish the processes of biotic and abiotic degradation of the substances evaluated, as well as environmental transformation through oxidation-reduction, biotransformation, and photochemical degradation, among others (Mitrano et al. 2015; Card et al. 2017). Thus, in this way, the antimicrobial activity must function as a primary screening to select the products that will be evaluated for toxicity in order to obtain a therapeutic index where the effective dose will be compared against the toxic dose of each compound (Leekha et al. 2011; Tamargo et al. 2015; Leontiev et al. 2018).

2.6 Therapeutic Index in Nanomaterials

To effectively correlate the toxic dose against the effective dose of a substance, it is important to establish its pharmacokinetics as well as its toxicokinetics, to understand how much it accumulates in the tissues and determine the possible damages it can cause in a correct therapeutic index (Reichard et al. 2016; Bell et al. 2018; Medina et al. 2018). Thus, the therapeutic index becomes a fundamental safety factor in the evaluation of nanomaterials and can be correlated using all toxicity parameters recommended by the OECD (Gao and Lowry 2018; Vitorino 2018; Zhu et al. 2019). Also, regarding environmental toxicity, an index should be obtained that determines the impact of nanomaterials on ecosystems; in which case the antimicrobial activity must be correlated with those obtained in battery of aquatic tests, as well as toxicity in plants (Bondarenko et al. 2016; Galdiero et al. 2016; Pedrazzani et al. 2019). Finally, this issue acquires fundamental importance in view of the great possibilities and challenges that nanotechnology imposes on us as a new tool for treatment, control, and diagnosis of diseases.

2.7 Conclusions

The emergence of antimicrobial resistance drives the development of new alternatives for obtaining new compounds and formulations with which infectious diseases can be treated and controlled, among which nanotechnology emerges as one of the most promising tools to face this threat to public health (Torres-Sangiao et al. 2016; Alvarez et al. 2017; Bilal et al. 2017). In this order of ideas, the methodologies acquire great importance in order to determine the antimicrobial activity of nanomaterials with great precision and reproducibility, as one of the strategies for obtaining nanoproducts that can translate into medicine, agriculture, and industry (Hartung and Sabbioni 2011; Hafner et al. 2014). Likewise, activity protocols must be correlated with toxicity assessment techniques to be able to select the most active products with the lowest toxicity to be applied in the different scenarios that antimicrobial resistance requires (Wesgate et al. 2016; LaCourse et al. 2018). Therefore, antimicrobial nanomaterials need correct protocols from translational science to reach an effective and safe application of products developed in the different circumstances where they are used (Hofmann-Amtenbrink et al. 2015; Ahonen et al. 2017; El-Sayed and Kamel 2019).

Acknowledgments The author thanks Sebastian Ritoré for his collaboration and invaluable support during the writing of this chapter, as well as the graphics contained in this book.

References

Adisa, I. O., Pullagurala, V. L. R., Peralta-Videa, J. R., Dimkpa, C. O., Elmer, W. H., Gardea-Torresdey, J., & White, J. (2019). Recent advances in nano-enabled fertilizers and pesticides: A critical review of mechanisms of action. *Environmental Science: Nano 6*, 2002–2030.

Adlhart, C., Verran, J., Azevedo, N. F., Olmez, H., Keinänen-Toivola, M. M., Gouveia, I., et al. (2018). Surface modifications for antimicrobial effects in the healthcare setting: A critical overview. *Journal of Hospital Infection, 99*(3), 239–249.

Ahonen, M., Kahru, A., Ivask, A., Kasemets, K., Kõljalg, S., Mantecca, P., et al. (2017). Proactive approach for safe use of antimicrobial coatings in healthcare settings: Opinion of the COST action network AMiCI. *International Journal of Environmental Research and Public Health, 14*(4), 366.

Akter, M., Sikder, M. T., Rahman, M. M., Ullah, A. A., Hossain, K. F. B., Banik, S., et al. (2018). A systematic review on silver nanoparticles-induced cytotoxicity: Physicochemical properties and perspectives. *Journal of Advanced Research, 9*, 1–16.

Allah, E. H., Saber, M., & Zaghloul, A. (2019). Nanotechnology applications in agriculture. *International Journal of Environmental Pollution and Environmental Modelling, 2*(4), 196–211.

Alvarez, M. M., Aizenberg, J., Analoui, M., Andrews, A. M., Bisker, G., Boyden, E. S., et al. (2017). Emerging trends in micro-and nanoscale technologies in medicine: From basic discoveries to translation. *ACS Nano, 11*(6), 5195.

Álvarez-Paino, M., Muñoz-Bonilla, A., & Fernández-García, M. (2017). Antimicrobial polymers in the nano-world. *Nanomaterials, 7*(2), 48.

Arévalo, L., Yarce, C., Oñate-Garzón, J., & Salamanca, C. (2019). Decrease of antimicrobial resistance through polyelectrolyte-coated nanoliposomes loaded with β-lactam drug. *Pharmaceuticals (Basel), 12*(1), 1.

Armas, F., Pacor, S., Ferrari, E., Guida, F., Pertinhez, T. A., Romani, A. A., et al. (2019). Design, antimicrobial activity and mechanism of action of Arg-rich ultra-short cationic lipopeptides. *PLoS One, 14*(2), e0212447.

Azam, A., Ahmed, A. S., Oves, M., Khan, M. S., Habib, S. S., & Memic, A. (2012). Antimicrobial activity of metal oxide nanoparticles against Gram-positive and Gram-negative bacteria: A comparative study. *International Journal of Nanomedicine, 7*, 6003.

Azeredo, J., Azevedo, N. F., Briandet, R., Cerca, N., Coenye, T., Costa, A. R., et al. (2017). Critical review on biofilm methods. *Critical Reviews in Microbiology, 43*(3), 313–351.

Balouiri, M., Sadiki, M., & Ibnsouda, S. K. (2016). Methods for in vitro evaluating antimicrobial activity: A review. *Journal of Pharmaceutical Analysis, 6*(2), 71–79.

Bell, S. M., Chang, X., Wambaugh, J. F., Allen, D. G., Bartels, M., Brouwer, K. L., et al. (2018). In vitro to in vivo extrapolation for high throughput prioritization and decision making. *Toxicology In Vitro, 47*, 213–227.

Bilal, M., Rasheed, T., Iqbal, H. M., Hu, H., Wang, W., & Zhang, X. (2017). Macromolecular agents with antimicrobial potentialities: A drive to combat antimicrobial resistance. *International Journal of Biological Macromolecules, 103*, 554–574.

Binas, V., Venieri, D., Kotzias, D., & Kiriakidis, G. (2017). Modified TiO2 based photocatalysts for improved air and health quality. *Journal of Materiomics, 3*(1), 3–16.

Bloomfield, S. F., Carling, P. C., & Exner, M. (2017). A unified framework for developing effective hygiene procedures for hands, environmental surfaces and laundry in healthcare, domestic, food handling and other settings. *GMS Hygiene and Infection Control, 12* Doc08.

Bogachev, M. I., Volkov, V. Y., Markelov, O. A., Trizna, E. Y., Baydamshina, D. R., Melnikov, V., et al. (2018). Fast and simple tool for the quantification of biofilm-embedded cells subpopulations from fluorescent microscopic images. *PLoS One, 13*(5), e0193267.

Bondarenko, O. M., Heinlaan, M., Sihtmäe, M., Ivask, A., Kurvet, I., Joonas, E., et al. (2016). Multilaboratory evaluation of 15 bioassays for (eco) toxicity screening and hazard ranking of engineered nanomaterials: FP7 project NANOVALID. *Nanotoxicology, 10*(9), 1229–1242.

Botha, T. L., Elemike, E. E., Horn, S., Onwudiwe, D. C., Giesy, J. P., & Wepener, V. (2019). Cytotoxicity of Ag, Au and Ag-Au bimetallic nanoparticles prepared using golden rod (Solidago canadensis) plant extract. *Scientific Reports, 9*(1), 4169.

Boudarel, H., Mathias, J. D., Blaysat, B., & Grédiac, M. (2018). Towards standardized mechanical characterization of microbial biofilms: Analysis and critical review. *NPJ Biofilms and Microbiomes, 4*(1), 1–15.

Bueno, J. (2014). Anti-biofilm drug susceptibility testing methods: Looking for new strategies against resistance mechanism. *Journal of Microbial & Biochemical Technology, 3*, 2.

Burduşel, A. C., Gherasim, O., Grumezescu, A. M., Mogoantă, L., Ficai, A., & Andronescu, E. (2018). Biomedical applications of silver nanoparticles: An up-to-date overview. *Nanomaterials, 8*(9), 681.

Card, M. L., Gomez-Alvarez, V., Lee, W. H., Lynch, D. G., Orentas, N. S., Lee, M. T., et al. (2017). History of EPI Suite™ and future perspectives on chemical property estimation in US Toxic Substances Control Act new chemical risk assessments. *Environmental Science: Processes and Impacts, 19*(3), 203–212.

Cardano, F., Frasconi, M., & Giordani, S. (2018). Photo-responsive graphene and carbon nanotubes to control and tackle biological systems. *Frontiers in Chemistry, 6*, 102.

Cattò, C., & Cappitelli, F. (2019). Testing anti-biofilm polymeric surfaces: Where to start? *International Journal of Molecular Sciences, 20*(15), 3794.

Chattopadhyay, P., Banerjee, G., & Mukherjee, S. (2017). Recent trends of modern bacterial insecticides for pest control practice in integrated crop management system. *3 Biotech, 7*(1), 60.

Chiu, L., Bazin, T., Truchetet, M. E., Schaeverbeke, T., Delhaes, L., & Pradeu, T. (2017). Protective microbiota: From localized to long-reaching co-immunity. *Frontiers in Immunology, 8*, 1678.

Chouhan, S., Sharma, K., & Guleria, S. (2017). Antimicrobial activity of some essential oils—present status and future perspectives. *Medicine, 4*(3), 58.

Cossarizza, A., Chang, H. D., Radbruch, A., Akdis, M., Andrä, I., Annunziato, F., et al. (2017). Guidelines for the use of flow cytometry and cell sorting in immunological studies. *European Journal of Immunology, 47*(10), 1584–1797.

de Melo Carrasco, L., Sampaio, J., & Carmona-Ribeiro, A. (2015). Supramolecular cationic assemblies against multidrug-resistant microorganisms: Activity and mechanism of action. *International Journal of Molecular Sciences, 16*(3), 6337–6352.

de Souza, I. O., Schrekker, C. M., Lopes, W., Orru, R. V., Hranjec, M., Perin, N., et al. (2016). Bifunctional fluorescent benzimidazo [1, 2-α] quinolines for Candida spp. biofilm detection and biocidal activity. *Journal of Photochemistry and Photobiology B: Biology, 163*, 319–326.

Dearfield, K. L., Gollapudi, B. B., Bemis, J. C., Benz, R. D., Douglas, G. R., Elespuru, R. K., et al. (2017). Next generation testing strategy for assessment of genomic damage: A conceptual framework and considerations. *Environmental and Molecular Mutagenesis, 58*(5), 264–283.

Deshmukh, S. P., Patil, S. M., Mullani, S. B., & Delekar, S. D. (2019). Silver nanoparticles as an effective disinfectant: A review. *Materials Science and Engineering. C. Materials for Biological Applications, 97*, 954.

Dijck, P. V., Sjollema, J., Cammue, B. P., Lagrou, K., Berman, J., d'Enfert, C., et al. (2018). Methodologies for in vitro and in vivo evaluation of efficacy of antifungal and antibiofilm agents and surface coatings against fungal biofilms. *Microbial Cell, 5*(7), 300.

Drasler, B., Sayre, P., Steinhaeuser, K. G., Petri-Fink, A., & Rothen-Rutishauser, B. (2017). In vitro approaches to assess the hazard of nanomaterials. *NanoImpact, 8*, 99–116.

Egorova, K. S., Gordeev, E. G., & Ananikov, V. P. (2017). Biological activity of ionic liquids and their application in pharmaceutics and medicine. *Chemical Reviews, 117*(10), 7132–7189.

Eloff, J. N. (2019). Avoiding pitfalls in determining antimicrobial activity of plant extracts and publishing the results. *BMC Complementary and Alternative Medicine, 19*(1), 106.

El-Sayed, A., & Kamel, M. (2019). Advances in nanomedical applications: Diagnostic, therapeutic, immunization, and vaccine production. *Environmental Science and Pollution Research 2019*, 1–14.

Emerson, J. B., Adams, R. I., Román, C. M. B., Brooks, B., Coil, D. A., Dahlhausen, K., et al. (2017). Schrödinger's microbes: Tools for distinguishing the living from the dead in microbial ecosystems. *Microbiome, 5*(1), 86.

Erdem, S. S., Khan, S., Palanisami, A., & Hasan, T. (2014). Rapid, low-cost fluorescent assay of β-lactamase-derived antibiotic resistance and related antibiotic susceptibility. *Journal of Biomedical Optics, 19*(10), 105007.

Fadeel, B., Bussy, C., Merino, S., Vázquez, E., Flahaut, E., Mouchet, F., et al. (2018). Safety assessment of graphene-based materials: Focus on human health and the environment. *ACS Nano, 12*(11), 10582–10620.

Francolini, I., Vuotto, C., Piozzi, A., & Donelli, G. (2017). Antifouling and antimicrobial biomaterials: An overview. *APMIS, 125*(4), 392–417.

Galdiero, E., Siciliano, A., Maselli, V., Gesuele, R., Guida, M., Fulgione, D., et al. (2016). An integrated study on antimicrobial activity and ecotoxicity of quantum dots and quantum dots coated with the antimicrobial peptide indolicidin. *International Journal of Nanomedicine, 11*, 4199.

Gao, X., & Lowry, G. V. (2018). Progress towards standardized and validated characterizations for measuring physicochemical properties of manufactured nanomaterials relevant to nano health and safety risks. *NanoImpact, 9*, 14–30.

Garcia, E., Shinde, R., Martinez, S., Kaushik, A., Chand, H. S., Nair, M., & Jayant, R. D. (2019). Cell-line-based studies of nanotechnology drug-delivery systems: A brief review. In *Nanocarriers for drug delivery* (pp. 375–393). Elsevier, Amsterdam, Netherlands.

Gazzola, G., Habimana, O., Murphy, C. D., & Casey, E. (2015). Comparison of biomass detachment from biofilms of two different Pseudomonas spp. under constant shear conditions. *Biofouling, 31*(1), 13–18.

Hafner, A., Lovrić, J., Lakoš, G. P., & Pepić, I. (2014). Nanotherapeutics in the EU: An overview on current state and future directions. *International Journal of Nanomedicine, 9*, 1005.

Han, G., & Ceilley, R. (2017). Chronic wound healing: A review of current management and treatments. *Advances in Therapy, 34*(3), 599–610.

Han, J., Zhao, D., Li, D., Wang, X., Jin, Z., & Zhao, K. (2018). Polymer-based nanomaterials and applications for vaccines and drugs. *Polymers, 10*(1), 31.

Hanson, M. A., Dostalova, A., Ceroni, C., Poidevin, M., Kondo, S., & Lemaitre, B. (2019). Synergy and remarkable specificity of antimicrobial peptides in vivo using a systematic knockout approach. *eLife, 8*, e44341.

Hartung, T., & Sabbioni, E. (2011). Alternative in vitro assays in nanomaterial toxicology. *Wiley Interdisciplinary Reviews: Nanomedicine and Nanobiotechnology, 3*(6), 545–573.

Helmi, K., David, F., Di Martino, P., Jaffrezic, M. P., & Ingrand, V. (2018). Assessment of flow cytometry for microbial water quality monitoring in cooling tower water and oxidizing biocide treatment efficiency. *Journal of Microbiological Methods, 152*, 201–209.

Hemeg, H. A. (2017). Nanomaterials for alternative antibacterial therapy. *International Journal of Nanomedicine, 12*, 8211.

Hofmann-Amtenbrink, M., Grainger, D. W., & Hofmann, H. (2015). Nanoparticles in medicine: Current challenges facing inorganic nanoparticle toxicity assessments and standardizations. *Nanomedicine: Nanotechnology, Biology and Medicine, 11*(7), 1689–1694.

Horká, M., Kubesová, A., Moravcová, D., Šalplachta, J., Šesták, J., Tesařová, M., & Růžička, F. (2016). Identification of nosocomial pathogens and antimicrobials using phenotypic techniques. *Frontiers in Clinical Drug Research: Anti-Infectives, 2*, 151.

Horky, P., Skalickova, S., Baholet, D., & Skladanka, J. (2018). Nanoparticles as a solution for eliminating the risk of mycotoxins. *Nanomaterials, 8*(9), 727.

Jamil, B., & Imran, M. (2018). Factors pivotal for designing of nanoantimicrobials: An exposition. *Critical Reviews in Microbiology, 44*(1), 79–94.

Jeevanandam, J., Barhoum, A., Chan, Y. S., Dufresne, A., & Danquah, M. K. (2018). Review on nanoparticles and nanostructured materials: History, sources, toxicity and regulations. *Beilstein Journal of Nanotechnology, 9*(1), 1050–1074.

Kadiyala, U., Kotov, N. A., & VanEpps, J. S. (2018). Antibacterial metal oxide nanoparticles: Challenges in interpreting the literature. *Current Pharmaceutical Design, 24*(8), 896–903.

Kamaruzzaman, N. F., Tan, L. P., Hamdan, R. H., Choong, S. S., Wong, W. K., Gibson, A. J., et al. (2019). Antimicrobial polymers: The potential replacement of existing antibiotics? *International Journal of Molecular Sciences, 20*(11), 2747.

Kerstens, M., Boulet, G., Clais, S., Lanckacker, E., Delputte, P., Maes, L., & Cos, P. (2015). A flow cytometric approach to quantify biofilms. *Folia Microbiologica, 60*(4), 335–342.

Khan, I., Saeed, K., & Khan, I. (2019). Nanoparticles: Properties, applications and toxicities. *Arabian Journal of Chemistry, 12*(7), 908–931.

Koo, H., Allan, R. N., Howlin, R. P., Stoodley, P., & Hall-Stoodley, L. (2017). Targeting microbial biofilms: Current and prospective therapeutic strategies. *Nature Reviews Microbiology, 15*(12), 740.

Kostakioti, M., Hadjifrangiskou, M., & Hultgren, S. J. (2013). Bacterial biofilms: Development, dispersal, and therapeutic strategies in the dawn of the postantibiotic era. *Cold Spring Harbor Perspectives in Medicine, 3*(4), a010306.

Kralik, P., & Ricchi, M. (2017). A basic guide to real time PCR in microbial diagnostics: Definitions, parameters, and everything. *Frontiers in Microbiology, 8*, 108.

Kramer, A., Dissemond, J., Kim, S., Willy, C., Mayer, D., Papke, R., et al. (2018). Consensus on wound antisepsis: Update 2018. *Skin Pharmacology and Physiology, 31*(1), 28–58.

Kumar, A., Kumar, P., Anandan, A., Fernandes, T. F., Ayoko, G. A., & Biskos, G. (2014). Engineered nanomaterials: Knowledge gaps in fate, exposure, toxicity, and future directions. *Journal of Nanomaterials, 2014*, 5.

LaCourse, K. D., Peterson, S. B., Kulasekara, H. D., Radey, M. C., Kim, J., & Mougous, J. D. (2018). Conditional toxicity and synergy drive diversity among antibacterial effectors. *Nature Microbiology, 3*(4), 440.

Larimer, C., Winder, E., Jeters, R., Prowant, M., Nettleship, I., Addleman, R. S., & Bonheyo, G. T. (2016). A method for rapid quantitative assessment of biofilms with biomolecular staining and image analysis. *Analytical and Bioanalytical Chemistry, 408*(3), 999–1008.

Leekha, S., Terrell, C. L., & Edson, R. S. (2011, February). General principles of antimicrobial therapy. In *Mayo Clinic proceedings* (Vol. 86, No. 2, pp. 156–167). Elsevier.

Leonard, H., Colodner, R., Halachmi, S., & Segal, E. (2018). Recent advances in the race to design a rapid diagnostic test for antimicrobial resistance. *ACS Sensors, 3*(11), 2202–2217.

Leontiev, R., Hohaus, N., Jacob, C., Gruhlke, M. C., & Slusarenko, A. J. (2018). A comparison of the antibacterial and antifungal activities of thiosulfinate analogues of allicin. *Scientific Reports, 8*(1), 6763.

Luo, J., Dong, B., Wang, K., Cai, S., Liu, T., Cheng, X., et al. (2017). Baicalin inhibits biofilm formation, attenuates the quorum sensing-controlled virulence and enhances Pseudomonas aeruginosa clearance in a mouse peritoneal implant infection model. *PLoS One, 12*(4), e0176883.

Magana, M., Sereti, C., Ioannidis, A., Mitchell, C. A., Ball, A. R., Magiorkinis, E., et al. (2018). Options and limitations in clinical investigation of bacterial biofilms. *Clinical Microbiology Reviews, 31*(3), e00084–e00016.

Makowski, M., Silva, Í. C., Pais do Amaral, C., Gonçalves, S., & Santos, N. C. (2019). Advances in lipid and metal nanoparticles for antimicrobial peptide delivery. *Pharmaceutics, 11*(11), 588.

Malone, M., Goeres, D. M., Gosbell, I., Vickery, K., Jensen, S., & Stoodley, P. (2017). Approaches to biofilm-associated infections: The need for standardized and relevant biofilm methods for clinical applications. *Expert Review of Anti-Infective Therapy, 15*(2), 147–156.

Martin-Serrano, Á., Gómez, R., Ortega, P., & de la Mata, F. J. (2019). Nanosystems as vehicles for the delivery of antimicrobial peptides (AMPs). *Pharmaceutics, 11*(9), 448.

Masters, E. A., Trombetta, R. P., de Mesy Bentley, K. L., Boyce, B. F., Gill, A. L., Gill, S. R., et al. (2019). Evolving concepts in bone infection: Redefining "biofilm", "acute vs. chronic osteomyelitis", "the immune proteome" and "local antibiotic therapy". *Bone Research, 7*(1), 1–18.

Medina, F. T., Andueza, I., & Suarez, A. I. (2018). Methods and protocols for in vivo animal nanotoxicity evaluation: A detailed review. In *Nanotoxicology* (pp. 323–388). CRC Press, Boca Raton, Florida.

Mitrano, D. M., Motellier, S., Clavaguera, S., & Nowack, B. (2015). Review of nanomaterial aging and transformations through the life cycle of nano-enhanced products. *Environment International, 77*, 132–147.

Mühlen, S., & Dersch, P. (2015). Anti-virulence strategies to target bacterial infections. In *How to overcome the antibiotic crisis* (pp. 147–183). Cham: Springer.

Munguia, J., & Nizet, V. (2017). Pharmacological targeting of the host–pathogen interaction: Alternatives to classical antibiotics to combat drug-resistant superbugs. *Trends in Pharmacological Sciences, 38*(5), 473–488.

O'Halloran, C., Walsh, N., O'Grady, M. C., Barry, L., Hooton, C., Corcoran, G. D., & Lucey, B. (2018). Assessment of the comparability of CLSI, EUCAST and stokes antimicrobial susceptibility profiles for Escherichia coli uropathogenic isolates. *British Journal of Biomedical Science, 75*(1), 24–29.

Panic, G., Flores, D., Ingram-Sieber, K., & Keiser, J. (2015). Fluorescence/luminescence-based markers for the assessment of Schistosoma mansoni schistosomula drug assays. *Parasites and Vectors, 8*(1), 624.

Panpaliya, N. P., Dahake, P. T., Kale, Y. J., Dadpe, M. V., Kendre, S. B., Siddiqi, A. G., & Maggavi, U. R. (2019). In vitro evaluation of antimicrobial property of silver nanoparticles and chlorhexidine against five different oral pathogenic bacteria. *The Saudi Dental Journal, 31*(1), 76–83.

Parmar, K. M., Hathi, Z. J., & Dafale, N. A. (2017). Control of multidrug-resistant gene flow in the environment through bacteriophage intervention. *Applied Biochemistry and Biotechnology, 181*(3), 1007–1029.

Patra, J. K., Das, G., Fraceto, L. F., Campos, E. V. R., del Pilar Rodriguez-Torres, M., Acosta-Torres, L. S., et al. (2018). Nano based drug delivery systems: Recent developments and future prospects. *Journal of Nanobiotechnology, 16*(1), 71.

Pavoni, L., Pavela, R., Cespi, M., Bonacucina, G., Maggi, F., Zeni, V., et al. (2019). Green micro- and nanoemulsions for managing parasites, vectors and pests. *Nanomaterials, 9*(9), 1285.

Pedrazzani, R., Bertanza, G., Brnardić, I., Cetecioglu, Z., Dries, J., Dvarionienė, J., et al. (2019). Opinion paper about organic trace pollutants in wastewater: Toxicity assessment in a European perspective. *Science of the Total Environment, 651*, 3202–3221.

Petersen, E. J., Diamond, S. A., Kennedy, A. J., Goss, G. G., Ho, K., Lead, J., et al. (2015). Adapting OECD aquatic toxicity tests for use with manufactured nanomaterials: Key issues and consensus recommendations. *Environmental Science and Technology, 49*(16), 9532–9547.

Pezzi, L., Pane, A., Annesi, F., Losso, M. A., Guglielmelli, A., Umeton, C., & De Sio, L. (2019). Antimicrobial effects of chemically functionalized and/or photo-heated nanoparticles. *Materials (Basel), 12*(7), 1078.

Raghunath, A., & Perumal, E. (2017). Metal oxide nanoparticles as antimicrobial agents: A promise for the future. *International Journal of Antimicrobial Agents, 49*(2), 137–152.

Rai, M., Deshmukh, S. D., Ingle, A. P., Gupta, I. R., Galdiero, M., & Galdiero, S. (2016). Metal nanoparticles: The protective nanoshield against virus infection. *Critical Reviews in Microbiology, 42*(1), 46–56.

Rana, S., & Kalaichelvan, P. T. (2013). Ecotoxicity of nanoparticles. *ISRN Toxicology, 2013*, 1.

Rancoita, P. M., Cugnata, F., Cruz, A. L. G., Borroni, E., Hoosdally, S. J., Walker, T. M., et al. (2018). Validating a 14-drug microtiter plate containing bedaquiline and delamanid for large-scale research susceptibility testing of Mycobacterium tuberculosis. *Antimicrobial Agents and Chemotherapy, 62*(9), e00344–e00318.

Reichard, J. F., Maier, M. A., Naumann, B. D., Pecquet, A. M., Pfister, T., Sandhu, R., et al. (2016). Toxicokinetic and toxicodynamic considerations when deriving health-based exposure limits for pharmaceuticals. *Regulatory Toxicology and Pharmacology, 79*, S67–S78.

Riediker, M., Zink, D., Kreyling, W., Oberdörster, G., Elder, A., Graham, U., et al. (2019). Particle toxicology and health-where are we? *Particle and Fibre Toxicology, 16*(1), 19.

Ristić, T., Persin, Z., Kralj Kuncic, M., Kosalec, I., & Zemljic, L. F. (2019). The evaluation of the in vitro antimicrobial properties of fibers functionalized by chitosan nanoparticles. *Textile Research Journal, 89*(5), 748–761.

Roy, R., Tiwari, M., Donelli, G., & Tiwari, V. (2018). Strategies for combating bacterial biofilms: A focus on anti-biofilm agents and their mechanisms of action. *Virulence, 9*(1), 522–554.

Ruddaraju, L. K., Pammi, S. V. N., Padavala, V. S., & Kolapalli, V. R. M. (2020). A review on anti-bacterials to combat resistance: From ancient era of plants and metals to present and future perspectives of green nano technological combinations. *Asian Journal of Pharmaceutical Sciences, 15*(1), 42–59.

Ruffin, M., & Brochiero, E. (2019). Repair process impairment by Pseudomonas aeruginosa in epithelial tissues: Major features and potential therapeutic avenues. *Frontiers in Cellular and Infection Microbiology, 9*, 182.

Sabaté Brescó, M., Harris, L. G., Thompson, K., Stanic, B., Morgenstern, M., O'Mahony, L., et al. (2017). Pathogenic mechanisms and host interactions in Staphylococcus epidermidis device-related infection. *Frontiers in Microbiology, 8*, 1401.

Sampath Kumar, T. S., & Madhumathi, K. (2014). Antibacterial potential of nanobioceramics used as drug carriers. In *Handbook of bioceramics and biocomposites* (pp. 1–42), Berlin/Heidelberg, Germany.

Savage, D. T., Hilt, J. Z., & Dziubla, T. D. (2019). In vitro methods for assessing nanoparticle toxicity. In *Nanotoxicity* (pp. 1–29). New York: Humana Press.

Sedghizadeh, P. P., Sun, S., Junka, A. F., Richard, E., Sadrerafi, K., Mahabady, S., et al. (2017). Design, synthesis, and antimicrobial evaluation of a novel bone-targeting bisphosphonate-ciprofloxacin conjugate for the treatment of osteomyelitis biofilms. *Journal of Medicinal Chemistry, 60*(6), 2326–2343.

Sharma, D., Misba, L., & Khan, A. U. (2019). Antibiotics versus biofilm: An emerging battleground in microbial communities. *Antimicrobial Resistance and Infection Control, 8*(1), 76.

Siegrist, S., Cörek, E., Detampel, P., Sandström, J., Wick, P., & Huwyler, J. (2019). Preclinical hazard evaluation strategy for nanomedicines. *Nanotoxicology, 13*(1), 73–99.

Singh, R. P., Choi, J. W., Tiwari, A., & Pandey, A. C. (2014). Functional nanomaterials for multifarious nanomedicine. In *Biosensors nanotechnology* (pp. 141–197), Wiley, Hoboken, New Jersey.

Sirelkhatim, A., Mahmud, S., Seeni, A., Kaus, N. H. M., Ann, L. C., Bakhori, S. K. M., et al. (2015). Review on zinc oxide nanoparticles: Antibacterial activity and toxicity mechanism. *Nano-Micro Letters, 7*(3), 219–242.

Soeteman-Hernandez, L. G., Apostolova, M. D., Bekker, C., Dekkers, S., Grafström, R. C., Groenewold, M., et al. (2019). Safe innovation approach: Towards an agile system for dealing with innovations. *Materials Today Communications, 20*, 100548.

Soltani, A. M., & Pouypouy, H. (2019). Standardization and regulations of nanotechnology and recent government policies across the world on nanomaterials. In *Advances in phytonanotechnology* (pp. 419–446). Academic Press, Cambridge, Massachusetts.

Stratton, C. W. (2018). Advanced phenotypic antimicrobial susceptibility testing methods. In *Advanced techniques in diagnostic microbiology* (pp. 69–98). Cham: Springer.

Taghipour, S., Hosseini, S. M., & Ataie-Ashtiani, B. (2019). Engineering nanomaterials for water and wastewater treatment: Review of classifications, properties and applications. *New Journal of Chemistry., 43*, 7902.

Tamargo, J., Le Heuzey, J. Y., & Mabo, P. (2015). Narrow therapeutic index drugs: A clinical pharmacological consideration to flecainide. *European Journal of Clinical Pharmacology, 71*(5), 549–567.

Thiruvengadam, M., Rajakumar, G., & Chung, I. M. (2018). Nanotechnology: Current uses and future applications in the food industry. *3 Biotech, 8*(1), 74.

Tidwell, T. J., De Paula, R., Smadi, M. Y., & Keasler, V. V. (2015). Flow cytometry as a tool for oilfield biocide efficacy testing and monitoring. *International Biodeterioration and Biodegradation, 98*, 26–34.

Torres, N. S., Montelongo-Jauregui, D., Abercrombie, J. J., Srinivasan, A., Lopez-Ribot, J. L., Ramasubramanian, A. K., & Leung, K. P. (2018). Antimicrobial and antibiofilm activity of

synergistic combinations of a commercially available small compound library with colistin against Pseudomonas aeruginosa. *Frontiers in Microbiology, 9*, 2541.
Torres-Sangiao, E., Holban, A. M., & Gestal, M. C. (2016). Advanced nanobiomaterials: Vaccines, diagnosis and treatment of infectious diseases. *Molecules, 21*(7), 867.
Vanhauteghem, D., Audenaert, K., Demeyere, K., Hoogendoorn, F., Janssens, G. P., & Meyer, E. (2019). Flow cytometry, a powerful novel tool to rapidly assess bacterial viability in metal working fluids: Proof-of-principle. *PLoS One, 14*(2), e0211583.
Vega-Jiménez, A. L., Vázquez-Olmos, A. R., Acosta-Gío, E., & Álvarez-Pérez, M. A. (2019). In vitro antimicrobial activity evaluation of metal oxide nanoparticles. In *Nanoemulsions-properties, fabrications and applications*. IntechOpen, Rijeka - Croatia.
Vimbela, G. V., Ngo, S. M., Fraze, C., Yang, L., & Stout, D. A. (2017). Antibacterial properties and toxicity from metallic nanomaterials. *International Journal of Nanomedicine, 12*, 3941.
Vitorino, C. V. (2018). Nanomedicine: Principles, properties and regulatory issues. *Frontiers in Chemistry, 6*, 360.
Von Borowski, R. G., Gnoatto, S. C. B., Macedo, A. J., & Gillet, R. (2018). Promising antibiofilm activity of peptidomimetics. *Frontiers in Microbiology, 9*, 2157.
Wang, Y., & Salazar, J. K. (2016). Culture-independent rapid detection methods for bacterial pathogens and toxins in food matrices. *Comprehensive Reviews in Food Science and Food Safety, 15*(1), 183–205.
Wang, L., Hu, C., & Shao, L. (2017). The antimicrobial activity of nanoparticles: Present situation and prospects for the future. *International Journal of Nanomedicine, 12*, 1227.
Wesgate, R., Grasha, P., & Maillard, J. Y. (2016). Use of a predictive protocol to measure the antimicrobial resistance risks associated with biocidal product usage. *American Journal of Infection Control, 44*(4), 458–464.
Wolfram, J., Zhu, M., Yang, Y., Shen, J., Gentile, E., Paolino, D., et al. (2015). Safety of nanoparticles in medicine. *Current Drug Targets, 16*(14), 1671–1681.
Yang, K., Han, Q., Chen, B., Zheng, Y., Zhang, K., Li, Q., & Wang, J. (2018). Antimicrobial hydrogels: Promising materials for medical application. *International Journal of Nanomedicine, 13*, 2217.
Yang, B., Chen, Y., & Shi, J. (2019). Reactive oxygen species (ROS)-based nanomedicine. *Chemical Reviews, 119*(8), 4881–4985.
Yildirimer, L., Thanh, N. T., Loizidou, M., & Seifalian, A. M. (2011). Toxicology and clinical potential of nanoparticles. *Nano Today, 6*(6), 585–607.
Yu, Y. J., Wang, X. H., & Fan, G. C. (2018). Versatile effects of bacterium-released membrane vesicles on mammalian cells and infectious/inflammatory diseases. *Acta Pharmacologica Sinica, 39*(4), 514.
Yusof, H. M., Mohamad, R., & Zaidan, U. H. (2019). Microbial synthesis of zinc oxide nanoparticles and their potential application as an antimicrobial agent and a feed supplement in animal industry: A review. *Journal of Animal Science and Biotechnology, 10*(1), 57.
Zeng, Q., Zhu, Y., Yu, B., Sun, Y., Ding, X., Xu, C., et al. (2018). Antimicrobial and antifouling polymeric agents for surface functionalization of medical implants. *Biomacromolecules, 19*(7), 2805–2811.
Zhu, S., Gong, L., Li, Y., Xu, H., Gu, Z., & Zhao, Y. (2019). Safety assessment of nanomaterials to eyes: An important but neglected issue. *Advanced Science*, 1802289.
Zoffmann, S., Vercruysse, M., Benmansour, F., Maunz, A., Wolf, L., Marti, R. B., et al. (2019). Machine learning-powered antibiotics phenotypic drug discovery. *Scientific Reports, 9*(1), 5013.

Chapter 3
Synergy and Antagonism: The Criteria of the Formulation

The stages of human development are to strive for:
Besitz [Possession]
Wissen [Knowledge]
Können [Ability]
Sein [Being]
—Erwin Schrödinger (1887–1961)

The more precise the measurement of position, the more imprecise the measurement of momentum, and vice versa
Every experiment destroys some of the knowledge of the system which was obtained by previous experiments
There is a fundamental error in separating the parts from the whole, the mistake of atomizing what should not be atomized.
Unity and complementarity constitute reality
—Werner Heisenberg (1901–1976)

Abstract The pharmaceutical formulation is the process by which the active molecule is combined to develop a final drug. Currently, in antimicrobial therapy, it is considered that the modulation of the formulation may be the key to the development of novel treatment strategies, because it can enhance antibiotic activity and allow the correct absorption and distribution of the drug in infected tissues. Thus, in that order of ideas, the search for synergism in the formulation using nanomaterials as a new tool in the development of new medicines requires different evaluation and design models in order to obtain medicines with adjuvant capacity and to have several mechanisms of antimicrobial action in a multi-target therapy. For this reason, the objective of this chapter is to give the conceptual and integrating elements of the modern therapeutic strategies that are being implemented to enhance the anti-infective pharmacopoeia, as well as evidence the application and evaluation models of nanotechnology in the search for new hybrid medicines capable of treating the patient without developing antibiotic resistance.

J. Bueno, *Preclinical Evaluation of Antimicrobial Nanodrugs*, Nanotechnology in the Life Sciences, https://doi.org/10.1007/978-3-030-43855-5_3

3.1 Introduction

The success of combining the diagnosis together with the antimicrobial medication in a theranostics approach will require determining the synergism of the formulation as a whole for the treatment of infectious disease (Zhu et al. 2014; Cheesman et al. 2017; Kevadiya et al. 2019; Van Giau et al. 2019). In this order of ideas, it is necessary to design screening platforms capable of selecting the combination of antimicrobial compounds to be tested depending on their synergism, antagonism, and adjuvant potential (Singh and Yeh 2017; Mgbeahuruike et al. 2019; Yilancioglu 2019). Likewise, possible drug interactions that alter the correct pharmacokinetics of antimicrobial drugs should also be evaluated, because this alters the plasma concentrations of the drug and predisposes the appearance of antimicrobial resistance (Guarino 2016; Boothe 2017; Yılmaz and Özcengiz 2017; Lee et al. 2019). Thus, in this order of ideas, it is necessary to correlate the anti-infective activity in vitro of the combinations of the different compounds with nanomaterials with the possible interactions and toxicity that may appear in order to increase the range of action and safety (Dakal et al. 2016; Ramón-García et al. 2016; Egorova and Ananikov 2017; Prasad et al. 2018). It is in this way that the screening platforms must be integrated into an in silico model that can predict the medications that can be combined with nanomaterials in an effective antimicrobial mixture, with good pharmacokinetics and that allows to increase the cure rates (Roncaglioni et al. 2013; Ventola 2017; Slikker Jr et al. 2018; Lombardo et al. 2019). Thus, in this order of ideas, the aim of this chapter is to analyze the fundamentals of the synergistic formulation in nanotechnology and how it can be applied in multi-target antimicrobial therapy to increase antibiotic potency safely (Baptista et al. 2018; Kumar et al. 2018; Lu et al. 2018; Xie and Xie 2019).

3.2 Multi-target Antimicrobial Strategy

Multi-target antimicrobial therapy consists in inhibiting with a compound or formulation several therapeutic targets within the microorganisms, such as the virulence and induction mechanisms of antimicrobial resistance (Fig. 3.1) (Gill et al. 2015; Koo et al. 2017; Baker et al. 2018; Liu et al. 2019). Likewise, the objective of multi-target therapy is to prevent the microorganism from invading the host cell and can activate the SOS response mechanism against antimicrobial therapy, which would lead to the microbial cell not acquiring any resistance (Culyba et al. 2015; Qin et al. 2015; Crane et al. 2018; Radlinski and Conlon 2018). Nanomaterials also have the ability to attack specific therapeutic targets in microorganisms such as the wall and cell membrane causing respiratory chain disruption (Beyth et al. 2015; Hemeg 2017; Gold et al. 2018; Yañez-Macías et al. 2019). Similarly, nanomaterials such as metal nanoparticles have shown activity as inhibitors of efflux pumps that expel antibiotics, as well as quorum sensing inhibitors, which is the set of chemical signals that drive biofilm formation (Gupta et al. 2017; Ahmad et al. 2019; Mantravadi et al. 2019). This means that nanoformulations can be antimicrobial

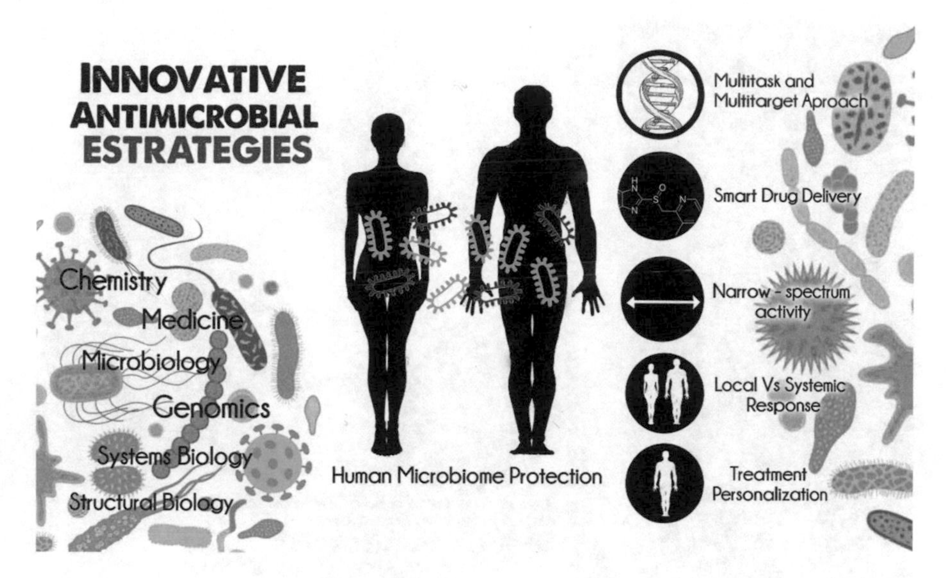

Fig. 3.1 Multi-target antimicrobial strategies

adjuvants, and, together with existing antibiotic medications, the possibility of designing and developing new antimicrobials capable of preventing the appearance of antimicrobial resistance is opened (Domalaon et al. 2018; Gupta and Datta 2019; Laws et al. 2019; Lima et al. 2019). Thus, obtaining hybrid antibiotics is a multi-target strategy that requires the implementation of synergism evaluation platforms that evaluate the inhibition of microbial therapeutic targets based on anti-infective activity, and the adjuvant activity on virulence and resistance mechanisms will be part of the new evaluation models (Pieren and Tigges 2012; Melander and Melander 2017; Tyers and Wright 2019; Wencewicz 2019). In this way, therapeutic targets that are synergistic with each other should be selected to establish potent antibiotic combinations, including cell wall, plasma membrane, respiratory chain, metabolic pathways, DNA, and protein synthesis, as well as efflux pumps, quorum sensing, SOS response, and horizontal transmission of genetic material (Fair and Tor 2014; Schwechheimer and Kuehn 2015; Trastoy et al. 2018; Baquero et al. 2019). Equally important is the design of the formulation that allows the correct absorption and distribution of the hybrid antimicrobial in the tissues; in this sense, the nanotransporters are a very useful tool that will allow the development of future therapies (Patra et al. 2018; Boyd et al. 2019; Canaparo et al. 2019; Lombardo et al. 2019).

3.3 Antimicrobial Nanoformulations in Drug Delivery

The modern nanoformulations allow to increase the absorption and distribution in the tissues of the antimicrobials without producing toxicity; for that reason they have become an element of development for new antibiotics (Chellan and Sadler 2015;

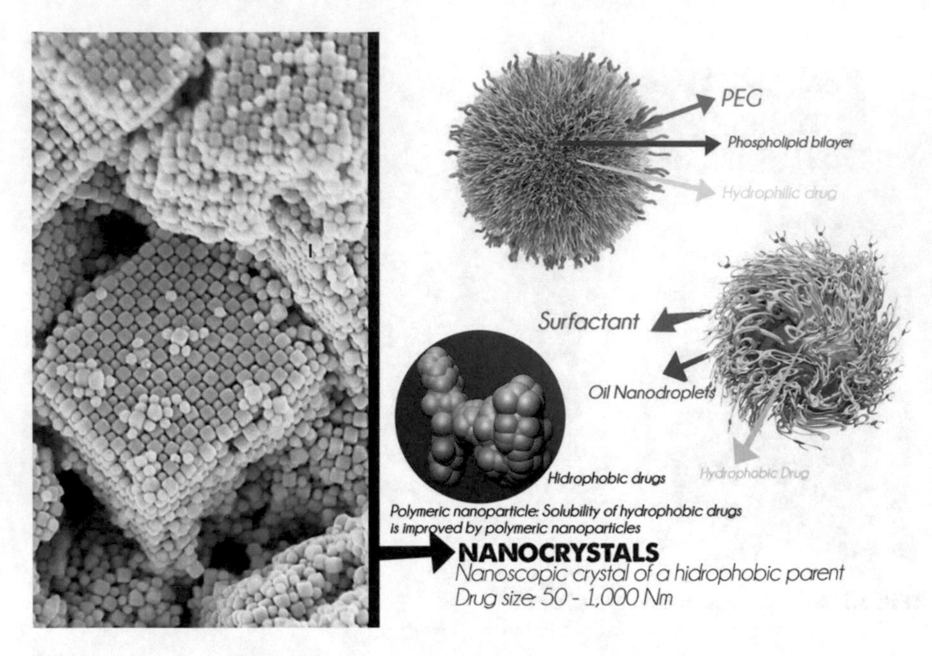

Fig. 3.2 Structure of antimicrobial nanoformulations

Wenzler et al. 2016; Teirlinck et al. 2018; Varier et al. 2019). Among these nanoformulations are liposomes, micelles, nanoemulsions, and polymeric nanoparticles, which can be functionalized with different bioactive molecules such as antibodies or anchor proteins to more efficiently transport antimicrobials to tissues affected by infectious disease (Fig. 3.2) (Mishra et al. 2018; Singh et al. 2019; Mehtani et al. 2019; Vasile 2019).

3.4 Antimicrobial Nanotheranostics Trojan Horse

One of the most promising applications of nanotheranostics is its use as a "Trojan horse" to help antibiotics enter into infected tissues and penetrate the less permeable microbial biofilms (Fig. 3.3) (Montanari et al. 2014; Xie et al. 2014; Agbale et al. 2016; Dosekova et al. 2017). In this way the siderophores nanomaterials are entered into the microorganism and with them drag antibiotics attached to their nanostructure and thus maintaining the intracellular inhibitory concentrations with which to destroy the microbial cells (Gupta et al. 2019; Shaikh et al. 2019). In this order of ideas, nanomaterials are not recognized by microorganisms as microbicidal agents and can enter by evading microbial defenses such as antibiotic-destroying enzymes and efflux pumps; it also maintains the integrity of the anti-infective inside the host until reaching therapeutic doses in tissues, allowing an era of renewal for old formulations that can be recognized and degraded by multiresistant microorganisms

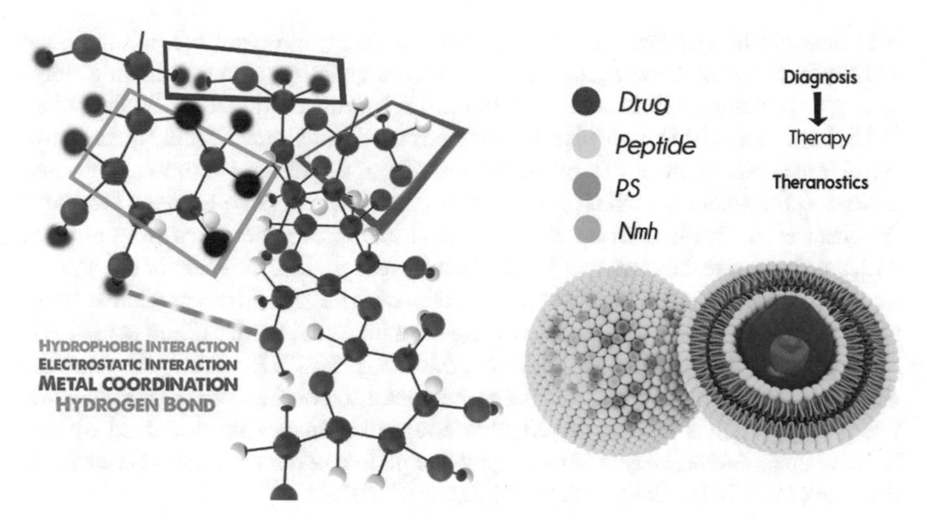

Fig. 3.3 Antimicrobial nanotheranostics

(Beloin et al. 2014; Mashitah et al. 2016; Santos et al. 2018; Yang et al. 2018). Thus, the Trojan horse effect of nanomaterials has to be evaluated in vitro for its synergy and toxicity, because in this strategy, the antimicrobial activity of the nanocomposite that enters the antibiotic should not be present so as not to trigger the adaptive responses of the microorganism (Aruguete et al. 2013; You et al. 2018; Azharuddin et al. 2019). In that order of ideas, nanoparticles with fluorescent markers can be used to determine the intracellular concentration of the nanomaterial in the microbial cell (Hemmerich and von Mikecz 2013; Maysinger et al. 2015; Samanta and Medintz 2016; Drasler et al. 2017). Thus, flow cytometry may be one of the most appropriate methodologies for determining the intracellular absorption and concentration of nanoparticles in both prokaryotic and eukaryotic cells (Araniti et al. 2018; Jiménez-Lamana et al. 2018; Tudose et al. 2019). In this way, by correlating the information on the cellular pharmacokinetics of nanomaterials in combination with antimicrobials, it is possible to design nanomedicaments with the effect of the Trojan horse that open the doors to the development of new formulations (Hsiao et al. 2015; Hoshyar et al. 2016).

3.5 Cellomics of Synergism Under a Multi-target Strategy

In this way, to determine the multi-target synergistic activity of a nanoformulation, it is necessary to use methods capable of quantifying the interaction and intracellular presence of the compounds to be evaluated; in this aspect, cellomics is the most appropriate approach for an in vitro study (Kang et al. 2017; Tam et al. 2018; Chernov et al. 2019; Yang et al. 2019). Thus, with these methods of image acquisition, such as flow cytometry, multiparametric analysis, and fluorescence microscopy, it is possible to visualize in a high-content screening protocol the antimicrobial activity in different therapeutic targets, as well as quantify the intracellular absorption of nanomaterials (Edwards and Sklar 2015; Martinez et al. 2015; Bankier et al. 2018; Lage et al. 2018). Likewise, the methods used in cellomics for the quantitative cellular analysis allow to determine the toxicity of the different combinations and antibiotic formulations, which gives a greater predictive value to these bioassays (Buchser et al. 2014; O'Brien 2014; Sindelar 2019). Another interesting topic in which cellomics techniques will be of importance is in the evaluation of the absorption of nanoformulations through Caco-2 cells, which is an in vitro predictive factor to determine if the nanodrug is able to cross the intestinal barrier (Zeng et al. 2012; Song et al. 2013; Brayden et al. 2015; Behzadi et al. 2017). Similarly, nanomaterial clearance concentration in tissues is another factor to consider when multi-targeted therapy approaches are developed, this concentration can be predicted by cell toxicity quantitative image analysis inside the patient under treatment (Desai 2012; Bartelink et al. 2019; Cova et al. 2019; Zhu et al. 2019).

3.6 Biofilm Penetration: Synergy and Strategy

Biofilms constitute a highly resistant microbial organization that allows microorganisms to persist and exchange genetic information for survival, so it is considered an association present both in nature and in infectious disease favoring invasion (Ramos et al. 2018; Sime-Ngando et al. 2018; Trastoy et al. 2018). It is for this reason that it has been sought to enhance antimicrobial formulations with nanomaterials capable of permeating biofilm barriers and preventing the consolidation of these niches of anti-infective resistance (Qayyum and Khan 2016; Vallet-Regí et al. 2019). In this route, fat-soluble nanomaterials have been able to enter the biofilms by increasing the concentrations of antimicrobials within the biological formation, inhibiting various therapeutic targets, and preventing the spread of persistent populations with epimutations (Piozzi and Francolini 2013; Galdiero et al. 2019). Similarly, several nanomaterials have shown the ability to dissolve the polysaccharide matrix that holds the biofilm together, which can lead to various antimicrobial compounds with microbicidal activity (Franci et al. 2015; Miller et al. 2015; Batoni et al. 2016). Thus, in this order of ideas, the synergy between antimicrobials becomes a phenomenon of adjuvant capacity that allows developing strategies to obtain a formulation with greater potency (Kalan and Wright 2011; Borges et al. 2016; Bueno 2016).

3.7 Conclusions

Thus, synergy not only consists in increasing the antimicrobial potency of a mixture or combination of compounds, but also the adjuvant capacity is an effective way to improve the activity of antimicrobial formulations, allowing the mechanisms of action and inhibition of anti-infective function in an optimal way (Álvarez-Paino et al. 2017; Borselli et al. 2019). Nanotechnology also offers the possibility of revitalizing all antimicrobial therapy that has lost activity with the emergence of multiresistance, which makes it a promising discipline in pharmacology (Shimanovich and Gedanken 2016). Because of the above, as important as the design and development of new antibiotic molecules is the integration of the compounds discovered in novel formulations that allow obtaining medications with greater curative capacity (Taylor 2015).

Acknowledgments The author thanks Sebastian Ritoré for his collaboration and invaluable support during the writing of this chapter, as well as the graphics contained in this book.

References

Agbale, C. M., Cardoso, M. H., Galyuon, I. K., & Franco, O. L. (2016). Designing metallodrugs with nuclease and protease activity. *Metallomics, 8*(11), 1159–1169.

Ahmad, I., Qais, F. A., Abulreesh, H. H., Ahmad, S., & Rumbaugh, K. P. (2019). Antibacterial drug discovery: Perspective insights. In *Antibacterial drug discovery to combat MDR* (pp. 1–21). Singapore: Springer.

Álvarez-Paino, M., Muñoz-Bonilla, A., & Fernández-García, M. (2017). Antimicrobial polymers in the nano-world. *Nanomaterials, 7*(2), 48.

Araniti, F., de la Peña, T. C., & Sánchez-Moreiras, A. M. (2018). Flow cytometric measurement of different physiological parameters. In *Advances in plant ecophysiology techniques* (pp. 195–213). Cham: Springer.

Aruguete, D. M., Kim, B., Hochella, M. F., Ma, Y., Cheng, Y., Hoegh, A., et al. (2013). Antimicrobial nanotechnology: Its potential for the effective management of microbial drug resistance and implications for research needs in microbial nanotoxicology. *Environmental Science: Processes and Impacts, 15*(1), 93–102.

Azharuddin, M., Zhu, G. H., Das, D., Ozgur, E., Uzun, L., Turner, A. P. F., & Patra, H. K. (2019). A repertoire of biomedical applications of noble metal nanoparticles. *Chemical communications (Cambridge, England), 55*(49), 6964–6996.

Baker, S. J., Payne, D. J., Rappuoli, R., & De Gregorio, E. (2018). Technologies to address antimicrobial resistance. *Proceedings of the National Academy of Sciences, 115*(51), 12887–12895.

Bankier, C., Cheong, Y., Mahalingam, S., Edirisinghe, M., Ren, G., Cloutman-Green, E., & Ciric, L. (2018). A comparison of methods to assess the antimicrobial activity of nanoparticle combinations on bacterial cells. *PLoS One, 13*(2), e0192093.

Baptista, P. V., McCusker, M. P., Carvalho, A., Ferreira, D. A., Mohan, N. M., Martins, M., & Fernandes, A. R. (2018). Nano-strategies to fight multidrug resistant bacteria-"A Battle of the Titans". *Frontiers in Microbiology, 9*, 1441.

Baquero, F., Lanza, V. F., Baquero, M. R., Del Campo, R., & Bravo-Vázquez, D. A. (2019). Microcins in Enterobacteriaceae: Peptide antimicrobials in the eco-active intestinal chemosphere. *Frontiers in Microbiology, 10*, 2261.

Bartelink, I. H., Jones, E. F., Shahidi-Latham, S. K., Lee, P. R. E., Zheng, Y., Vicini, P., et al. (2019). Tumor drug penetration measurements could be the neglected piece of the personalized cancer treatment puzzle. *Clinical Pharmacology and Therapeutics, 106*(1), 148–163.

Batoni, G., Maisetta, G., & Esin, S. (2016). Antimicrobial peptides and their interaction with biofilms of medically relevant bacteria. *Biochimica et Biophysica Acta (BBA)-Biomembranes, 1858*(5), 1044–1060.

Behzadi, S., Serpooshan, V., Tao, W., Hamaly, M. A., Alkawareek, M. Y., Dreaden, E. C., et al. (2017). Cellular uptake of nanoparticles: Journey inside the cell. *Chemical Society Reviews, 46*(14), 4218–4244.

Beloin, C., Renard, S., Ghigo, J. M., & Lebeaux, D. (2014). Novel approaches to combat bacterial biofilms. *Current Opinion in Pharmacology, 18*, 61–68.

Beyth, N., Houri-Haddad, Y., Domb, A., Khan, W., & Hazan, R. (2015). Alternative antimicrobial approach: Nano-antimicrobial materials. *Evidence-based Complementary and Alternative Medicine: eCAM, 2015*, 246012.

Boothe, D. M. (2017). Drug selection and dosing regimens. In *Monitoring and intervention for the critically ill small animal: The rule of 20* (pp. 319–332), Wiley, Hoboken, New Jersey.

Borges, A., Abreu, A. C., Dias, C., Saavedra, M. J., Borges, F., & Simões, M. (2016). New perspectives on the use of phytochemicals as an emergent strategy to control bacterial infections including biofilms. *Molecules, 21*(7), 877.

Borselli, D., Brunel, J. M., Gorgé, O., & Bolla, J. M. (2019). Polyamino-isoprenyl derivatives as antibiotic adjuvants and motility inhibitors for Bordetella bronchiseptica porcine pulmonary infection treatment. *Frontiers in Microbiology, 10*, 1771.

Boyd, B. J., Bergström, C. A., Vinarov, Z., Kuentz, M., Brouwers, J., Augustijns, P., et al. (2019). Successful oral delivery of poorly water-soluble drugs both depends on the intraluminal behavior of drugs and of appropriate advanced drug delivery systems. *European Journal of Pharmaceutical Sciences, 137*, 104967.

Brayden, D. J., Cryan, S. A., Dawson, K. A., O'Brien, P. J., & Simpson, J. C. (2015). High-content analysis for drug delivery and nanoparticle applications. *Drug Discovery Today, 20*(8), 942–957.

Buchser, W., Collins, M., Garyantes, T., Guha, R., Haney, S., Lemmon, V., et al. (2014). Assay development guidelines for image-based high content screening, high content analysis and high content imaging. In *Assay guidance manual [Internet]*. Eli Lilly and Company and the National Center for Advancing Translational Sciences. Bethesda, Maryland.

Bueno, J. (2016). Antimicrobial adjuvants drug discovery, the challenge of avoid the resistance and recover the susceptibility of multidrug-resistant strains. *Journal of Microbial & Biochemical Technology, 8*, 169–176.

Canaparo, R., Foglietta, F., Giuntini, F., Della Pepa, C., Dosio, F., & Serpe, L. (2019). Recent developments in antibacterial therapy: Focus on stimuli-responsive drug-delivery systems and therapeutic nanoparticles. *Molecules, 24*(10), 1991.

Cheesman, M. J., Ilanko, A., Blonk, B., & Cock, I. E. (2017). Developing new antimicrobial therapies: Are synergistic combinations of plant extracts/compounds with conventional antibiotics the solution? *Pharmacognosy Reviews, 11*(22), 57.

Chellan, P., & Sadler, P. J. (2015). The elements of life and medicines. *Philosophical Transactions of the Royal Society A: Mathematical, Physical and Engineering Sciences, 373*(2037), 20140182.

Chernov, V. M., Chernova, O. A., Mouzykantov, A. A., Lopukhov, L. L., & Aminov, R. I. (2019). Omics of antimicrobials and antimicrobial resistance. *Expert Opinion on Drug Discovery, 14*(5), 455–468.

Cova, T. F., Bento, D. J., & Nunes, S. C. (2019). Computational approaches in theranostics: Mining and predicting cancer data. *Pharmaceutics, 11*(3), 119.

Crane, J. K., Cheema, M. B., Olyer, M. A., & Sutton, M. D. (2018). Zinc blockade of SOS response inhibits horizontal transfer of antibiotic resistance genes in enteric bacteria. *Frontiers in Cellular and Infection Microbiology, 8*, 410.

Culyba, M. J., Mo, C. Y., & Kohli, R. M. (2015). Targets for combating the evolution of acquired antibiotic resistance. *Biochemistry, 54*(23), 3573–3582.

Dakal, T. C., Kumar, A., Majumdar, R. S., & Yadav, V. (2016). Mechanistic basis of antimicrobial actions of silver nanoparticles. *Frontiers in Microbiology, 7*, 1831.

Desai, N. (2012). Challenges in development of nanoparticle-based therapeutics. *The AAPS Journal, 14*(2), 282–295.

Domalaon, R., Idowu, T., Zhanel, G. G., & Schweizer, F. (2018). Antibiotic hybrids: The next generation of agents and adjuvants against gram-negative pathogens? *Clinical Microbiology Reviews, 31*(2), e00077–e00017.

Dosekova, E., Filip, J., Bertok, T., Both, P., Kasak, P., & Tkac, J. (2017). Nanotechnology in glycomics: Applications in diagnostics, therapy, imaging, and separation processes. *Medicinal Research Reviews, 37*(3), 514–626.

Drasler, B., Sayre, P., Steinhaeuser, K. G., Petri-Fink, A., & Rothen-Rutishauser, B. (2017). In vitro approaches to assess the hazard of nanomaterials. *NanoImpact, 8*, 99–116.

Edwards, B. S., & Sklar, L. A. (2015). Flow cytometry: Impact on early drug discovery. *Journal of Biomolecular Screening, 20*(6), 689–707.

Egorova, K. S., & Ananikov, V. P. (2017). Toxicity of metal compounds: Knowledge and myths. *Organometallics, 36*(21), 4071–4090.

Fair, R. J., & Tor, Y. (2014). Antibiotics and bacterial resistance in the 21st century. *Perspectives in Medicinal Chemistry, 6*, PMC-S14459.

Franci, G., Falanga, A., Galdiero, S., Palomba, L., Rai, M., Morelli, G., & Galdiero, M. (2015). Silver nanoparticles as potential antibacterial agents. *Molecules, 20*(5), 8856–8874.

Galdiero, E., Lombardi, L., Falanga, A., Libralato, G., Guida, M., & Carotenuto, R. (2019). Biofilms: Novel strategies based on antimicrobial peptides. *Pharmaceutics, 11*(7), 322.

Gill, E. E., Franco, O. L., & Hancock, R. E. (2015). Antibiotic adjuvants: Diverse strategies for controlling drug-resistant pathogens. *Chemical Biology and Drug Design, 85*(1), 56–78.

Gold, K., Slay, B., Knackstedt, M., & Gaharwar, A. K. (2018). Antimicrobial activity of metal and metal-oxide based nanoparticles. *Advanced Therapeutics, 1*(3), 1700033.

Guarino, R. A. (2016). Adverse events and reactions: Etiology, drug interactions, collection, and reporting. In *New drug approval process* (pp. 370–393). CRC Press. Boca Raton, Florida

Gupta, V., & Datta, P. (2019). Next-generation strategy for treating drug resistant bacteria: Antibiotic hybrids. *The Indian Journal of Medical Research, 149*(2), 97.

Gupta, D., Singh, A., & Khan, A. U. (2017). Nanoparticles as efflux pump and biofilm inhibitor to rejuvenate bactericidal effect of conventional antibiotics. *Nanoscale Research Letters, 12*(1), 1–6.

Gupta, A., Mumtaz, S., Li, C. H., Hussain, I., & Rotello, V. M. (2019). Combatting antibiotic-resistant bacteria using nanomaterials. *Chemical Society Reviews, 48*(2), 415–427.

Hemeg, H. A. (2017). Nanomaterials for alternative antibacterial therapy. *International Journal of Nanomedicine, 12*, 8211.

Hemmerich, P. H., & von Mikecz, A. H. (2013). Defining the subcellular interface of nanoparticles by live-cell imaging. *PLoS One, 8*(4), e62018.

Hoshyar, N., Gray, S., Han, H., & Bao, G. (2016). The effect of nanoparticle size on in vivo pharmacokinetics and cellular interaction. *Nanomedicine, 11*(6), 673–692.

Hsiao, I. L., Hsieh, Y. K., Wang, C. F., Chen, I. C., & Huang, Y. J. (2015). Trojan-horse mechanism in the cellular uptake of silver nanoparticles verified by direct intra-and extracellular silver speciation analysis. *Environmental Science and Technology, 49*(6), 3813–3821.

Jiménez-Lamana, J., Szpunar, J., & Łobinski, R. (2018). New frontiers of metallomics: Elemental and species-specific analysis and imaging of single cells. In *Metallomics* (pp. 245–270). Cham: Springer.

Kalan, L., & Wright, G. D. (2011). Antibiotic adjuvants: multicomponent anti-infective strategies. *Expert Reviews in Molecular Medicine, 13*.

Kang, T., Zhu, Q., Wei, D., Feng, J., Yao, J., Jiang, T., et al. (2017). Nanoparticles coated with neutrophil membranes can effectively treat cancer metastasis. *ACS Nano, 11*(2), 1397–1411.

Kevadiya, B. D., Ottemann, B. M., Thomas, M. B., Mukadam, I., Nigam, S., McMillan, J., et al. (2019). Neurotheranostics as personalized medicines. *Advanced Drug Delivery Reviews, 148*, 252–289.

Koo, H., Allan, R. N., Howlin, R. P., Stoodley, P., & Hall-Stoodley, L. (2017). Targeting microbial biofilms: Current and prospective therapeutic strategies. *Nature Reviews Microbiology, 15*(12), 740.

Kumar, M., Curtis, A., & Hoskins, C. (2018). Application of nanoparticle technologies in the combat against anti-microbial resistance. *Pharmaceutics, 10*(1), 11.

Lage, O. M., Ramos, M. C., Calisto, R., Almeida, E., Vasconcelos, V., & Vicente, F. (2018). Current screening methodologies in drug discovery for selected human diseases. *Marine Drugs, 16*(8), 279.

Laws, M., Shaaban, A., & Rahman, K. M. (2019). Antibiotic resistance breakers: Current approaches and future directions. *FEMS Microbiology Reviews, 43*(5), 490–516.

Lee, W., Cai, Y., Lim, T. P., Teo, J., Chua, S. C., & Kwa, A. L. H. (2019). In vitro pharmacodynamics and PK/PD in animals. In *Polymyxin antibiotics: From laboratory bench to bedside* (pp. 105–116). Cham: Springer.

Lima, R., Del Fiol, F. S., & Balcão, V. M. (2019). Prospects for the use of new technologies in combating multidrug-resistant bacteria. *Frontiers in Pharmacology, 10*, 692.

Liu, Y., Ding, S., Shen, J., & Zhu, K. (2019). Nonribosomal antibacterial peptides that target multidrug-resistant bacteria. *Natural Product Reports, 36*(4), 573–592.

Lombardo, D., Kiselev, M. A., & Caccamo, M. T. (2019). Smart nanoparticles for drug delivery application: Development of versatile nanocarrier platforms in biotechnology and nanomedicine. *Journal of Nanomaterials, 2019*, 1.

Lu, A., Zhang, C., Zhou, W., Guan, D., & Wang, Y. (2018). Network intervention: A new therapeutic strategy. *Frontiers in Pharmacology, 9*, 754.

Mantravadi, P. K., Kalesh, K. A., Dobson, R. C., Hudson, A. O., & Parthasarathy, A. (2019). The quest for novel antimicrobial compounds: Emerging trends in research, development, and technologies. *Antibiotics, 8*(1), 8.

Martinez, N. J., Titus, S. A., Wagner, A. K., & Simeonov, A. (2015). High-throughput fluorescence imaging approaches for drug discovery using in vitro and in vivo three-dimensional models. *Expert Opinion on Drug Discovery, 10*(12), 1347–1361.

Mashitah, M. D., San Chan, Y., & Jason, J. (2016). Antimicrobial properties of nanobiomaterials and the mechanism. In *Nanobiomaterials in antimicrobial therapy* (pp. 261–312). William Andrew Publishing. Boca Raton, Florida. Norwich, NY.

Maysinger, D., Ji, J., Hutter, E., & Cooper, E. (2015). Nanoparticle-based and bioengineered probes and sensors to detect physiological and pathological biomarkers in neural cells. *Frontiers in Neuroscience, 9*, 480.

Mehtani, D., Seth, A., Sharma, P., Maheshwari, N., Kapoor, D., Shrivastava, S. K., & Tekade, R. K. (2019). Biomaterials for sustained and controlled delivery of small drug molecules. In *Biomaterials and bionanotechnology* (pp. 89–152). Academic Press. Cambridge, Massachusetts.

Melander, R. J., & Melander, C. (2017). The challenge of overcoming antibiotic resistance: An adjuvant approach? *ACS Infectious Diseases, 3*(8), 559–563.

Mgbeahuruike, E. E., Stålnacke, M., Vuorela, H., & Holm, Y. (2019). Antimicrobial and synergistic effects of commercial piperine and piperlongumine in combination with conventional antimicrobials. *Antibiotics, 8*(2), 55.

Miller, K. P., Wang, L., Benicewicz, B. C., & Decho, A. W. (2015). Inorganic nanoparticles engineered to attack bacteria. *Chemical Society Reviews, 44*(21), 7787–7807.

Mishra, D. K., Shandilya, R., & Mishra, P. K. (2018). Lipid based nanocarriers: A translational perspective. *Nanomedicine: Nanotechnology, Biology and Medicine, 14*(7), 2023–2050.

Montanari, E., D'Arrigo, G., Di Meo, C., Virga, A., Coviello, T., Passariello, C., & Matricardi, P. (2014). Chasing bacteria within the cells using levofloxacin-loaded hyaluronic acid nanohydrogels. *European Journal of Pharmaceutics and Biopharmaceutics, 87*(3), 518–523.

O'Brien, P. J. (2014). High-content analysis in toxicology: Screening substances for human toxicity potential, elucidating subcellular mechanisms and in vivo use as translational safety biomarkers. *Basic and Clinical Pharmacology and Toxicology, 115*(1), 4–17.

Patra, J. K., Das, G., Fraceto, L. F., Campos, E. V. R., del Pilar Rodriguez-Torres, M., Acosta-Torres, L. S., et al. (2018). Nano based drug delivery systems: Recent developments and future prospects. *Journal of Nanobiotechnology, 16*(1), 71.

Pieren, M., & Tigges, M. (2012). Adjuvant strategies for potentiation of antibiotics to overcome antimicrobial resistance. *Current Opinion in Pharmacology, 12*(5), 551–555.

Piozzi, A., & Francolini, I. (2013). Editorial of the special issue antimicrobial polymers. *International Journal of Molecular Sciences, 14*(9), 18002.

Prasad, M., Lambe, U. P., Brar, B., Shah, I., Manimegalai, J., Ranjan, K., et al. (2018). Nanotherapeutics: An insight into healthcare and multi-dimensional applications in medical sector of the modern world. *Biomedicine and Pharmacotherapy, 97*, 1521–1537.

Qayyum, S., & Khan, A. U. (2016). Nanoparticles vs. biofilms: A battle against another paradigm of antibiotic resistance. *Medicinal Chemistry Communication, 7*(8), 1479–1498.

Qin, T. T., Kang, H. Q., Ma, P., Li, P. P., Huang, L. Y., & Gu, B. (2015). SOS response and its regulation on the fluoroquinolone resistance. *Annals of Translational Medicine, 3*(22), 358.

Radlinski, L., & Conlon, B. P. (2018). Antibiotic efficacy in the complex infection environment. *Current Opinion in Microbiology, 42*, 19–24.

Ramón-García, S., Del Río, R. G., Villarejo, A. S., Sweet, G. D., Cunningham, F., Barros, D., et al. (2016). Repurposing clinically approved cephalosporins for tuberculosis therapy. *Scientific Reports, 6*, 34293.

Ramos, M. A. D. S., Da Silva, P. B., Sposito, L., De Toledo, L. G., Bonifacio, B. V., Rodero, C. F., et al. (2018). Nanotechnology-based drug delivery systems for control of microbial biofilms: A review. *International Journal of Nanomedicine, 13*, 1179.

Roncaglioni, A., Toropov, A. A., Toropova, A. P., & Benfenati, E. (2013). In silico methods to predict drug toxicity. *Current Opinion in Pharmacology, 13*(5), 802–806.

Samanta, A., & Medintz, I. L. (2016). Nanoparticles and DNA–a powerful and growing functional combination in bionanotechnology. *Nanoscale, 8*(17), 9037–9095.

Santos, R. S., Figueiredo, C., Azevedo, N. F., Braeckmans, K., & De Smedt, S. C. (2018). Nanomaterials and molecular transporters to overcome the bacterial envelope barrier: Towards advanced delivery of antibiotics. *Advanced Drug Delivery Reviews, 136*, 28–48.

Schwechheimer, C., & Kuehn, M. J. (2015). Outer-membrane vesicles from Gram-negative bacteria: Biogenesis and functions. *Nature Reviews Microbiology, 13*(10), 605.

Shaikh, S., Nazam, N., Rizvi, S. M. D., Ahmad, K., Baig, M. H., Lee, E. J., & Choi, I. (2019). Mechanistic insights into the antimicrobial actions of metallic nanoparticles and their implications for multidrug resistance. *International Journal of Molecular Sciences, 20*(10), 2468.

Shimanovich, U., & Gedanken, A. (2016). Nanotechnology solutions to restore antibiotic activity. *Journal of Materials Chemistry B, 4*(5), 824–833.

Sime-Ngando, T., Bertrand, J. C., Bogusz, D., Brugère, J. F., Franche, C., Fardeau, M. L., et al. (2018). The evolution of living beings started with prokaryotes and in interaction with prokaryotes. In *Prokaryotes and evolution* (pp. 241–338). Cham: Springer.

Sindelar, R. D. (2019). Genomics, other "OMIC" technologies, precision medicine, and additional biotechnology-related techniques. In *Pharmaceutical biotechnology* (pp. 191–237). Cham: Springer.

Singh, N., & Yeh, P. J. (2017). Suppressive drug combinations and their potential to combat antibiotic resistance. *The Journal of Antibiotics, 70*(11), 1033.

Singh, A. K., Yadav, T. P., Pandey, B., Gupta, V., & Singh, S. P. (2019). Engineering nanomaterials for smart drug release: Recent advances and challenges. In *Applications of targeted nano drugs and delivery systems* (pp. 411–449). Elsevier. Amsterdam, Netherlands.

Slikker, W., Jr., de Souza Lima, T. A., Archella, D., de Silva Junior, J. B., Barton-Maclaren, T., Bo, L., et al. (2018). Emerging technologies for food and drug safety. *Regulatory Toxicology and Pharmacology, 98*, 115–128.

Song, Q., Wang, X., Hu, Q., Huang, M., Yao, L., Qi, H., et al. (2013). Cellular internalization pathway and transcellular transport of pegylated polyester nanoparticles in Caco-2 cells. *International Journal of Pharmaceutics, 445*(1–2), 58–68.

Tam, J., Hamza, T., Ma, B., Chen, K., Beilhartz, G. L., Ravel, J., et al. (2018). Host-targeted niclosamide inhibits C. difficile virulence and prevents disease in mice without disrupting the gut microbiota. *Nature Communications, 9*(1), 5233.

Taylor, D. (2015). The pharmaceutical industry and the future of drug development. In *Pharmaceuticals in the environment* (pp. 1–33).

Teirlinck, E., Xiong, R., Brans, T., Forier, K., Fraire, J., Van Acker, H., et al. (2018). Laser-induced vapour nanobubbles improve drug diffusion and efficiency in bacterial biofilms. *Nature Communications, 9*(1), 4518.

Trastoy, R., Manso, T., Fernandez-Garcia, L., Blasco, L., Ambroa, A., Del Molino, M. P., et al. (2018). Mechanisms of bacterial tolerance and persistence in the gastrointestinal and respiratory environments. *Clinical Microbiology Reviews, 31*(4), e00023–e00018.

Tudose, M., Culita, D. C., Voicescu, M., Musuc, A. M., Kuncser, A. C., Bleotu, C., et al. (2019). Fluorescent coumarin-modified mesoporous SBA-15 nanocomposite: Physico-chemical characterization and interaction with prokaryotic and eukaryotic cells. *Microporous and Mesoporous Materials, 288*, 109583.

Tyers, M., & Wright, G. D. (2019). Drug combinations: A strategy to extend the life of antibiotics in the 21st century. *Nature Reviews Microbiology, 17*(3), 141–155.

Vallet-Regí, M., González, B., & Izquierdo-Barba, I. (2019). Nanomaterials as promising alternative in the infection treatment. *International Journal of Molecular Sciences, 20*(15), 3806.

Van Giau, V., An, S. S. A., & Hulme, J. (2019). Recent advances in the treatment of pathogenic infections using antibiotics and nano-drug delivery vehicles. *Drug Design, Development and Therapy, 13*, 327.

Varier, K. M., Gudeppu, M., Chinnasamy, A., Thangarajan, S., Balasubramanian, J., Li, Y., & Gajendran, B. (2019). Nanoparticles: Antimicrobial applications and its prospects. In *Advanced nanostructured materials for environmental remediation* (pp. 321–355). Cham: Springer.

Vasile, C. (2019). Polymeric nanomaterials: Recent developments, properties and medical applications. In *Polymeric nanomaterials in nanotherapeutics* (pp. 1–66). Elsevier. Amsterdam, Netherlands.

Ventola, C. L. (2017). Progress in nanomedicine: Approved and investigational nanodrugs. *Pharmacy and Therapeutics, 42*(12), 742.

Wencewicz, T. A. (2019). Crossroads of antibiotic resistance and biosynthesis. *Journal of Molecular Biology, 431*, 3370.

Wenzler, E., Fraidenburg, D. R., Scardina, T., & Danziger, L. H. (2016). Inhaled antibiotics for Gram-negative respiratory infections. *Clinical Microbiology Reviews, 29*(3), 581–632.

Xie, L., & Xie, L. (2019). Pathway-centric structure-based multi-target compound screening for anti-virulence drug repurposing. *International Journal of Molecular Sciences, 20*(14), 3504.

Xie, S., Tao, Y., Pan, Y., Qu, W., Cheng, G., Huang, L., et al. (2014). Biodegradable nanoparticles for intracellular delivery of antimicrobial agents. *Journal of Controlled Release, 187*, 101–117.

Yañez-Macías, R., Muñoz-Bonilla, A., Jesús-Tellez, D., Marco, A., Maldonado-Textle, H., Guerrero-Sánchez, C., et al. (2019). Combinations of antimicrobial polymers with nanomaterials and bioactives to improve biocidal therapies. *Polymers, 11*(11), 1789.

Yang, K., Han, Q., Chen, B., Zheng, Y., Zhang, K., Li, Q., & Wang, J. (2018). Antimicrobial hydrogels: Promising materials for medical application. *International Journal of Nanomedicine, 13*, 2217.

Yang, C. Y., Hsu, C. Y., Fang, C. S., Shiau, C. W., Chen, C. S., & Chiu, H. C. (2019). Loxapine, an antipsychotic drug, suppresses intracellular multiple-antibiotic-resistant Salmonella enterica serovar Typhimurium in macrophages. *Journal of Microbiology, Immunology and Infection, 52*(4), 638–647.

Yilancioglu, K. (2019). Antimicrobial drug interactions: Systematic evaluation of protein and nucleic acid synthesis inhibitors. *Antibiotics, 8*(3), 114.

Yılmaz, Ç., & Özcengiz, G. (2017). Antibiotics: Pharmacokinetics, toxicity, resistance and multidrug efflux pumps. *Biochemical Pharmacology, 133*, 43–62.

You, F., Tang, W., & Yung, L. Y. L. (2018). Real-time monitoring of the Trojan-horse effect of silver nanoparticles by using a genetically encoded fluorescent cell sensor. *Nanoscale, 10*(16), 7726–7735.

Zeng, N., Gao, X., Hu, Q., Song, Q., Xia, H., Liu, Z., et al. (2012). Lipid-based liquid crystalline nanoparticles as oral drug delivery vehicles for poorly water-soluble drugs: Cellular interaction and in vivo absorption. *International Journal of Nanomedicine, 7*, 3703.
Zhu, X., Radovic-Moreno, A. F., Wu, J., Langer, R., & Shi, J. (2014). Nanomedicine in the management of microbial infection–overview and perspectives. *Nano Today, 9*(4), 478–498.
Zhu, X., Vo, C., Taylor, M., & Smith, B. R. (2019). Non-spherical micro-and nanoparticles in nanomedicine. *Materials Horizons, 6*, 1094.

Chapter 4
In Vitro Nanotoxicity: Toward the Development of Safe and Effective Treatments

The ten commandments according to Leó Szilárd1.
1. Recognize the connections of things and laws of conduct of men, so that you may know what you are doing.
2. Let your acts be directed toward a worthy goal, but do not ask if they will reach it; they are to be models and examples, not means to an end.
3. Speak to all men as you do to yourself, with no concern for the effect you make, so that you do not shut them out from your world; lest in isolation the meaning of life slips out of sight and you lose the belief in the perfection of creation.
4. Do not destroy what you cannot create.
5. Touch no dish, except that you are hungry.
6. Do not covet what you cannot have.
7. Do not lie without need.
8. Honor children. Listen reverently to their words and speak to them with infinite love.
9. Do your work for six years; but in the seventh, go into solitude or among strangers, so that the memory of your friends does not hinder you from being what you have become.
10. Lead your life with a gentle hand and be ready to leave whenever you are called.Leo Szilard (1898–1964) "Die Stimme der Delphine."
Utopische Erzählungen. Rowohit Taschenbuch Verlag. 1963.

Abstract One of the major limitations in the translation of the research results obtained in nanomaterial investigations is the presence of toxicity in organs and tissues. This toxicity can eventually lead to tissue damage, necrosis, and chronic inflammation, with potential induction of the phenomenon of carcinogenesis. Thus, it is a priority in any initiative that aims to design nanomedicaments to have adequate screening platforms to predict the toxicological behavior of the products obtained. Then, in this order of ideas, several aspects must be considered, depending on the use and mechanism of action of the nanocomposites, as well as the nanostructures, which must involve the risk, the correlation between the minimum effective and toxic doses, as well as the different possibilities to implement risk mitigation

J. Bueno, *Preclinical Evaluation of Antimicrobial Nanodrugs*, Nanotechnology in the Life Sciences, https://doi.org/10.1007/978-3-030-43855-5_4

strategies during therapy. For the above reasons, the objective of this chapter is to integrate the necessary concepts that are required in case you want to evaluate and determine the risk of toxicity of nanomaterials, with a view to establishing adequate safety parameters in the discovery of new antimicrobial drugs.

4.1 Introduction

The biggest concern caused by the massive use of nanomaterials is the possibility of accumulation in tissues, with the subsequent consequences of systemic toxicity (Ahmad et al. 2016; Alshehri et al. 2016; Wen et al. 2017; Fadeel et al. 2018). Nanotoxicity is a determining factor that prevents modern hybrid antimicrobial drugs with which to prevent the emergence of antimicrobial resistance from being used in the treatment of infectious diseases (Nikalje 2015; Lee et al. 2018; Joshi et al. 2019; Naskar and Kim 2019). Thus, the evaluation and identification of both drug and environmental toxicity should be a priority for the selection of promising nanomaterials to be used in the design and development of nanoantibiotics (Sahlgren et al. 2017; Sunderland et al. 2017; Vimbela et al. 2017; Ahmad et al. 2019). It is also very important to design new formulations that allow clearance of nanomaterials employees in order to be safely excreted (Kumar et al. 2013; Patra et al. 2018; Vitorino 2018; Navya et al. 2019). Equally necessary is to design modern evaluation platforms that determine and predict the possible interactions that nanopharmaceuticals may have with the tissues, as well as correctly establish the toxic concentrations of the compounds in order to be able to adjust the dosages properly (Williams et al. 2005; Onoue et al. 2014; Ventola 2017; Jeevanandam et al. 2018). In this order of ideas, the use of image-based strategies for the measurement of pharmacokinetics is a priority and thus determine in vivo the distribution and accumulation of nanocomposites in the body (Tuntland et al. 2014; Xing et al. 2014; Gobbo et al. 2015; Liu et al. 2018). Thus, in this therapeutic approach, it will be possible to evaluate the concentration in real time of the administered nanomedicines and to be able to give early toxicity alerts that can guide appropriate treatment protocols more safely (Mu et al. 2018; Rampado et al. 2019; Singh et al. 2019; Zottel et al. 2019). Therefore, the objective of this chapter will be to project the possibilities and integral solutions proposed by the challenge of evaluating nanotoxicity in an effective and early way (Zhang et al. 2012; Mourdikoudis et al. 2018; Qiu et al. 2018).

4.2 Nanomaterials in Interaction with Tissues

Some nanomaterials can destroy cell membranes and produce genotoxicity by their interaction with the nucleic acids of eukaryotic cells; they can also alter the mitochondrial respiratory chain and induce the production of reactive oxygen species

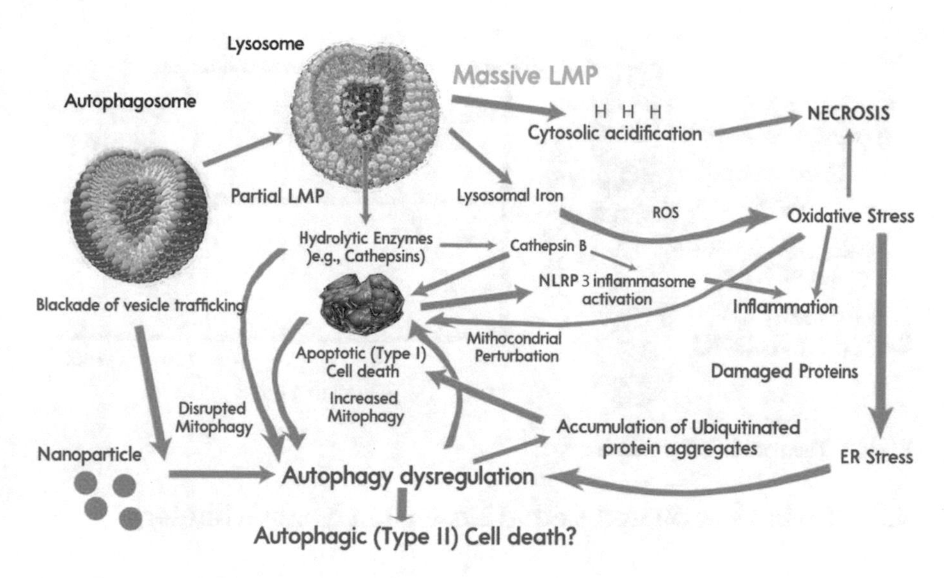

Fig. 4.1 Nanotoxicity mechanisms

(ROS) (Fig. 4.1) (Manke et al. 2013; Fu et al. 2014; Abdal Dayem et al. 2017; Paraskevaidi et al. 2017). Thus, nanocomposites are physical disruptors of plasma membranes, disrupting ion exchange and increasing their permeability, due to their ability to diffuse through lipid bilayers (Trimble and Grinstein 2015; Yang and Hinner 2015; Farnoud and Nazemidashtarjandi 2019). They can also produce epimutagenesis by altering the expression of certain specific genes for normal cell functioning (Sierra et al. 2016; Huang et al. 2017). Therefore, the possibility of interaction and toxicity of nanocomposites with host tissues is very high and requires a thorough evaluation prior to the design and development of these substances, as well as their combination with other molecules with confirmed toxicity (McCallion et al. 2016; Müller et al. 2017; Choudhury et al. 2019; Sharma et al. 2019). In this order of ideas, further studies should be carried out to reduce and prevent the induction of ROS by nanomaterials in normal host cells, either by increasing the specificity of nanomedicines or by binding them to antioxidant substances (Kurutas 2015; Tan et al. 2018; Yuan et al. 2018; Gao et al. 2019). For these reasons cytotoxicity is one of the consequences of the indiscriminate use of nanomaterials, so the toxic dose 50 (TD50) and the lethal dose 50 (LD50) for each nanocomposite to be used must be determined for each tissue in order to give an appropriate safety margin due to which this toxicity is dose dependent (Kong et al. 2011; Bahadar et al. 2016; Sahu et al. 2016).

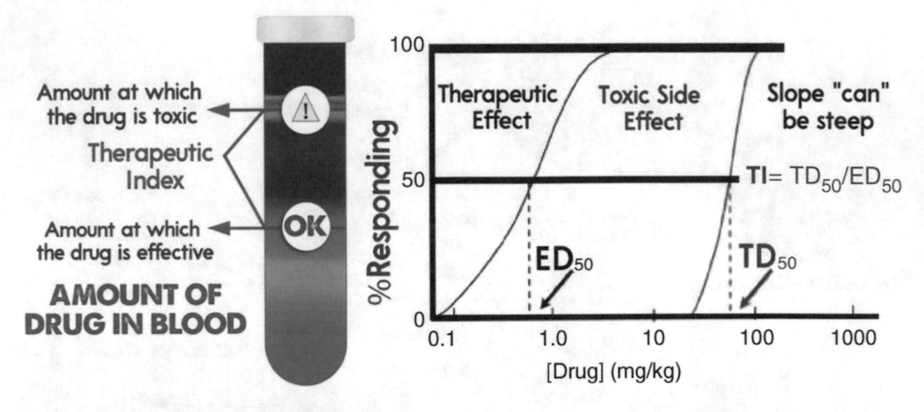

Fig. 4.2 Therapeutic index parameters

4.3 Toxic Dose 50 and Lethal Dose 50 in Nanotechnology

TD50 and LD50 should always be above the minimum effective dose (MED) and the maximum tolerated dose (MTD) for any pharmacologically active molecule, in order to obtain an adequate therapeutic index (Fig. 4.2) (Chou 2006; Tardiff and Rodricks 2013; Aston et al. 2017). The therapeutic index is a predicting value that determines if both a compound and molecule can be promising to become a successful drug in the treatment of a disease, since it correlates activity with toxicity (Kramer et al. 2007; Kashif et al. 2015; Tamargo et al. 2015; Bulusu et al. 2016). In this order of ideas, the therapeutic index of nanomaterials becomes a factor to be determined and improved, as well as the combinations developed with them to obtain hybrid medicines (Jurj et al. 2017; Gurunathan et al. 2018; Rizvi and Saleh 2018; Sang et al. 2019). In this way, a TD50 and LD50 determined both in cells and in vivo models and with organisms that live in terrestrial and aquatic environments are necessary to establish the true toxic potential of nanocomposites, as well as to evaluate their wide use in medicine and industry (Chinedu et al. 2013; Kleandrova et al. 2015; Ettrup et al. 2017; Erhirhie et al. 2018). Similarly, drug interactions between nanomaterials and drugs should be determined in order to avoid toxicity and adverse events as a result of improper combination during clinical treatment (Ferdousi et al. 2017; ud Din et al. 2017; Navya et al. 2019; Lin et al. 2019).

4.4 Drug Interactions and Adverse Events with Nanodrugs

Drug interactions and their adverse events are the product of the synergistic cytotoxicity present when combining one or more medications during clinical treatment (Yu et al. 2015; Gupta et al. 2017; Gerber et al. 2018; Chudzik et al. 2019). In this way, these adverse reactions in nanoformulations are produced by organic toxicity or immune reaction, which triggers systemic failure with prolonged use

(Casals et al. 2017; De Matteis 2017; Yuan et al. 2019; Lombardo et al. 2019). Thus, immunotoxicity becomes the greatest risk of the use of nanomaterials for the development of new medications, due to the possibility of inducing generalized hypersensitivity, cell destruction, and necrosis (Elsabahy and Wooley 2013; Wolfram et al. 2015; Dusinska et al. 2017; Drasler et al. 2017). For this reason it is necessary to determine the release of cytokines, as well as the immunomodulation induced by the nanopharmaceuticals that trigger the cascade of effects that make up the immune response (Dobrovolskaia 2015; Halamoda-Kenzaoui and Bremer-Hoffmann 2018; La-Beck et al. 2019; Lim et al. 2019). Likewise, immunotoxicity by inducing a prolonged immune response produces genotoxicity and epi-genotoxicity in the patient under treatment, which makes it a risk to consider when implementing a nanodrug development program (Sharma et al. 2012; Jiao et al. 2014; Tavares et al. 2014; Engin and Hayes 2018).

4.5 Immunotoxicity: Immune Response Without Control

The immunotoxicity of nanocomposites is a factor of concern due to the possibility of inducing cellular damage with indiscriminate use, as well as the risk of producing unwanted autoimmune reactions in patients (Dietert et al. 2010; Senapati et al. 2015; Smolkova et al. 2017; Bawa et al. 2019). Likewise, autoimmune reactions limit the application of nanotheranostics approaches to disease control and diagnosis, which makes its presence a predictor of the therapeutic index (Spivak et al. 2013; Clemente-Casares and Santamaria 2014; Sonali et al. 2018; Babele et al. 2019). Likewise, the presence of the nanoparticle-associated protein bio-corona, which is largely made up of human serum albumin (HSA), immunoglobulin (IgG), and fibrinogen, makes the use of

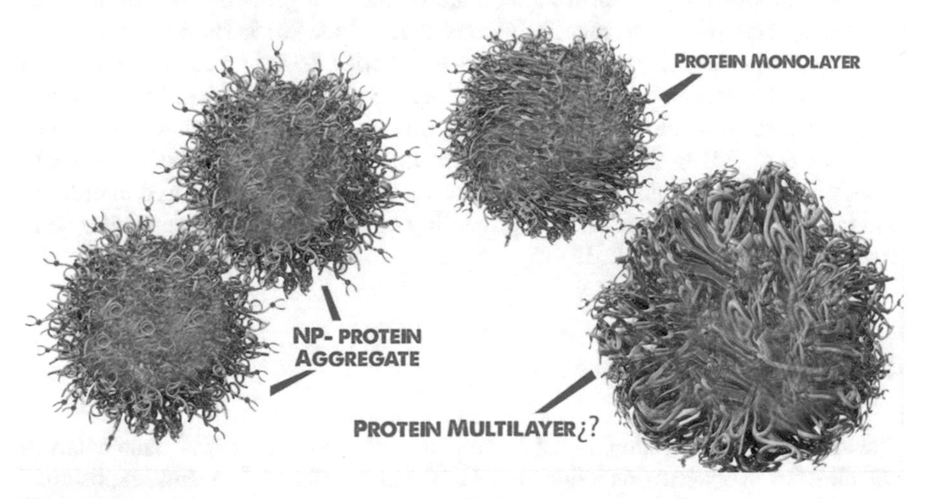

Fig. 4.3 Protein corona formation with nanomaterials

nanomedicaments increase the risk of immunotoxicity (Fig. 4.3) (Cristian et al. 2019). On the other hand, there is a risk that nanomaterials increase the rate of the appearance of immune cell neoplasms due to genotoxicity derived (Borm et al. 2006; Åkerlund et al. 2019; Li et al. 2019; Senchukova 2019). In this case, the possibility of carcinogenesis caused by oxidative stress produced by materials derived from nanotechnology applications deserves further study to avoid the complications of this approach for the treatment of infectious disease (Federico et al. 2007; Yu et al. 2008; Dekkers et al. 2016; Yazdimamaghani et al. 2018; Dickinson et al. 2019). Thus, the carcinogenicity of nanomaterials as a result of chronic inflammation and oxidative stress is a safety issue that must be solved as more treatment possibilities are evaluated (Oberdörster et al. 2005; Buzea et al. 2007; Ray et al. 2009; Seaton et al. 2009).

4.6 Genotoxicity, Epi-genotoxicity, and Carcinogenesis

In this vein, genotoxicity and carcinogenesis which induced both ROS can determine the success or failure of the clinical use of nanomedicine (Guo et al. 2011; Manke et al. 2013; Kwon et al. 2014; Alarifi et al. 2017). This is the reason why the presence of genotoxicity and carcinogenicity should be controlled and evaluated in nanoformulations that combine both drugs and diagnostic biomarkers (Ventola 2017; Roberti et al. 2019). Thus, this dose-dependent toxicity of nanomaterials should be predicted in its cost-benefit for patients who require it, depending on the organs affected and the individual pharmacogenetic profile (Davis et al. 2010; Fornaguera and García-Celma 2017; Simeonidis et al. 2019). Likewise, the epi-genotoxicity as well as its correlation with the pharmacogenomics should be determined when administering personalized nanomedicines (Tajbakhsh 2011; Thomson et al. 2012).

On the other hand, it is very important to correlate delayed hypersensitivity reactions as a predisposing factor for nanomaterial-induced genotoxicity with the corresponding risk of carcinogenesis (Goodson III et al. 2015; Hardy et al. 2018). Likewise, hypersensitivity phenomena have a strong genetic predisposition, so a personalized precision nanomedicine requires recognizing and identifying patients susceptible to presenting this type of reaction in the presence of nanomaterials (Ganguly et al. 2018). Thus, the chemical carcinogenesis associated with nanotechnology requires determining the toxicokinetics of the formulations used in order to predict tissue accumulation and the possibility of developing adverse events (Laux et al. 2018; Riediker et al. 2019).

4.7 Nanotoxicokinetics

The absorption, distribution, metabolism, and excretion (ADME) of nanomaterials will allow us to obtain a predictive model of the toxic effects of systemic exposure to nanopharmacology (Radomska et al. 2016; Cheah et al. 2017). Thus, this approach is important to develop a dose-response curve of the tissues against the nanocomposites and thus determine the correlation between the toxic dose and the effective dose (Tsatsakis et al. 2018). Nanotheranostics will also be of paramount support to know the distribution and accumulation in real time of the nanoparticles in the tissues of the patient under treatment, because it can be coupled to the fluorescent marker nanostructures that allow to determine the tissue concentration (Arms et al. 2018; Morales-Dalmau et al. 2019). In this order of ideas, nanotoxicokinetics is a necessary transdisciplinary approach to the design and development of new nanotechnological drugs through the synthesis processes (Marchant et al. 2008; Gilbertson et al. 2015). Thus, in the implementation of any program to discover new applications of nanomaterials, it is a priority to harmonize the nanosafety and nanotoxicology protocols in order to offer products with a high impact on health, but with fewer adverse reactions (Hobson et al. 2016; Warheit 2018). Finally, beyond the implementation of methodologies, it is of fundamental importance to apply a comprehensive toxicity assessment that allows the prediction of a successful therapy, which will always depend on a correct personalized correlation between the effective and the toxic dose (Krewski et al. 2010; Raies and Bajic 2016).

4.8 Conclusions

Safety in the application of engineered nanomaterials becomes the major limitation that currently has this promising approach (Teow et al. 2011; Schwarz-Plaschg et al. 2017). Therefore, it is necessary to record all the information on nanotoxicology both in vitro and in vivo in the different organs and tissues in order to develop predictive modeling platforms capable of determining the applications of nanomaterials in medicine (Nel et al. 2012; Neagu et al. 2017). Likewise, the design and development of nanostructures coupled to antioxidant molecules that increase tolerance to nanomaterials must be converted into an important line of research work in order to achieve an adequate therapeutic index (Prado-Audelo et al. 2019). On the other hand, nanotechnology requires a significant effort in translational medicine in toxicity in order to develop multidisciplinary approaches that impact modern pharmacopoeias (Lavik and von Recum 2011; Hua et al. 2018). Thus, adequate nanotherapy will depend on obtaining medications with very low minimum effective doses, which have a high cytoprotective capacity and with a metabolism and excretion in an allowed range (Jesus et al. 2019).

Acknowledgments The author thanks Sebastian Ritoré for his collaboration and invaluable support during the writing of this chapter, as well as the graphics contained in this book.

References

Abdal Dayem, A., Hossain, M. K., Lee, S. B., Kim, K., Saha, S. K., Yang, G. M., et al. (2017). The role of reactive oxygen species (ROS) in the biological activities of metallic nanoparticles. *International Journal of Molecular Sciences, 18*(1), 120.

Ahmad, M. Z., Abdel-Wahab, B. A., Alam, A., Zafar, S., Ahmad, J., Ahmad, F. J., et al. (2016). Toxicity of inorganic nanoparticles used in targeted drug delivery and other biomedical application: An updated account on concern of biomedical nanotoxicology. *Journal of Nanoscience and Nanotechnology, 16*(8), 7873–7897.

Ahmad, I., Qais, F. A., Abulreesh, H. H., Ahmad, S., & Rumbaugh, K. P. (2019). Antibacterial drug discovery: Perspective insights. In *Antibacterial drug discovery to combat MDR* (pp. 1–21). Singapore: Springer.

Åkerlund, E., Islam, M. S., McCarrick, S., Alfaro-Moreno, E., & Karlsson, H. L. (2019). Inflammation and (secondary) genotoxicity of Ni and NiO nanoparticles. *Nanotoxicology, 13*(8), 1060–1072.

Alarifi, S., Ali, D., Alkahtani, S., & Almeer, R. S. (2017). ROS-mediated apoptosis and genotoxicity induced by palladium nanoparticles in human skin malignant melanoma cells. *Oxidative Medicine and Cellular Longevity, 2017*, 8439098.

Alshehri, R., Ilyas, A. M., Hasan, A., Arnaout, A., Ahmed, F., & Memic, A. (2016). Carbon nanotubes in biomedical applications: Factors, mechanisms, and remedies of toxicity: Miniperspective. *Journal of Medicinal Chemistry, 59*(18), 8149–8167.

Arms, L., Smith, D. W., Flynn, J., Palmer, W., Martin, A., Woldu, A., & Hua, S. (2018). Advantages and limitations of current techniques for analyzing the biodistribution of nanoparticles. *Frontiers in Pharmacology, 9*, 802.

Aston, W. J., Hope, D. E., Nowak, A. K., Robinson, B. W., Lake, R. A., & Lesterhuis, W. J. (2017). A systematic investigation of the maximum tolerated dose of cytotoxic chemotherapy with and without supportive care in mice. *BMC Cancer, 17*(1), 684.

Babele, P. K., Gedda, M. R., Zahra, K., & Madhukar, P. (2019). Epigenetic aspects of engineered nanomaterials: Is the collateral damage inevitable? *Frontiers in Bioengineering and Biotechnology, 7*, 228.

Bahadar, H., Maqbool, F., Niaz, K., & Abdollahi, M. (2016). Toxicity of nanoparticles and an overview of current experimental models. *Iranian Biomedical Journal, 20*(1), 1.

Bawa, R., Szebeni, J., Webster, T. J., & Audette, G. F. (2019). *Immune aspects of biopharmaceuticals and nanomedicines*. Milton: Pan Stanford.

Borm, P. J., Robbins, D., Haubold, S., Kuhlbusch, T., Fissan, H., Donaldson, K., et al. (2006). The potential risks of nanomaterials: A review carried out for ECETOC. *Particle and Fibre Toxicology, 3*(1), 11.

Bulusu, K. C., Guha, R., Mason, D. J., Lewis, R. P., Muratov, E., Motamedi, Y. K., et al. (2016). Modelling of compound combination effects and applications to efficacy and toxicity: State-of-the-art, challenges and perspectives. *Drug Discovery Today, 21*(2), 225–238.

Buzea, C., Pacheco, I. I., & Robbie, K. (2007). Nanomaterials and nanoparticles: Sources and toxicity. *Biointerphases, 2*(4), MR17–MR71.

Casals, E., Gusta, M. F., Piella, J., Casals, G., Jiménez, W., & Puntes, V. (2017). Intrinsic and extrinsic properties affecting innate immune responses to nanoparticles: The case of cerium oxide. *Frontiers in Immunology, 8*, 970.

Cheah, H. Y., Kiew, L. V., Lee, H. B., Japundžić-Žigon, N., Vicent, M. J., Hoe, S. Z., & Chung, L. Y. (2017). Preclinical safety assessments of nano-sized constructs on cardiovascular system toxicity: A case for telemetry. *Journal of Applied Toxicology, 37*(11), 1268–1285.

Chinedu, E., Arome, D., & Ameh, F. S. (2013). A new method for determining acute toxicity in animal models. *Toxicology International, 20*(3), 224.

Chou, T. C. (2006). Theoretical basis, experimental design, and computerized simulation of synergism and antagonism in drug combination studies. *Pharmacological Reviews, 58*(3), 621–681.

Choudhury, P., Dinda, S., & Kumar, D. P. (2019). Fabrication of soft-nanocomposites from functional molecules with diversified applications. *Soft Matter, 16*, 27.

Chudzik, B., Bonio, K., Dabrowski, W., Pietrzak, D., Niewiadomy, A., Olender, A., et al. (2019). Synergistic antifungal interactions of amphotericin B with 4-(5-methyl-1, 3, 4-thiadiazole-2-yl) benzene-1, 3-diol. *Scientific Reports, 9*(1), 1–14.

Clemente-Casares, X., & Santamaria, P. (2014). Nanomedicine in autoimmunity. *Immunology Letters, 158*(1–2), 167–174.

Cristian, R. E., Mohammad, I. J., Mernea, M., Sbarcea, B. G., Trica, B., Stan, M. S., & Dinischiotu, A. (2019). Analyzing the interaction between two different types of nanoparticles and serum albumin. *Materials, 12*(19), 3183.

Davis, M. E., Chen, Z., & Shin, D. M. (2010). Nanoparticle therapeutics: An emerging treatment modality for cancer. In *Nanoscience and technology: A collection of reviews from nature journals* (pp. 239–250) World Scientific, Singapore.

De Matteis, V. (2017). Exposure to inorganic nanoparticles: Routes of entry, immune response, biodistribution and in vitro/in vivo toxicity evaluation. *Toxics, 5*(4), 29.

Dekkers, S., Oomen, A. G., Bleeker, E. A., Vandebriel, R. J., Micheletti, C., Cabellos, J., et al. (2016). Towards a nanospecific approach for risk assessment. *Regulatory Toxicology and Pharmacology, 80*, 46–59.

Dickinson, A. M., Godden, J. M., Lanovyk, K., & Ahmed, S. S. (2019). Assessing the safety of nanomedicines: A mini review. *Applied In Vitro Toxicology, 5*(3), 114–122.

Dietert, R. R., Dietert, J. M., & Gavalchin, J. (2010). Risk of autoimmune disease: Challenges for immunotoxicity testing. In *Immunotoxicity testing* (pp. 39–51). New York: Humana Press.

Dobrovolskaia, M. A. (2015). Pre-clinical immunotoxicity studies of nanotechnology-formulated drugs: Challenges, considerations and strategy. *Journal of Controlled Release, 220*, 571–583.

Drasler, B., Sayre, P., Steinhaeuser, K. G., Petri-Fink, A., & Rothen-Rutishauser, B. (2017). In vitro approaches to assess the hazard of nanomaterials. *NanoImpact, 8*, 99–116.

Dusinska, M., Tulinska, J., El Yamani, N., Kuricova, M., Liskova, A., Rollerova, E., et al. (2017). Immunotoxicity, genotoxicity and epigenetic toxicity of nanomaterials: New strategies for toxicity testing? *Food and Chemical Toxicology, 109*, 797–811.

Elsabahy, M., & Wooley, K. L. (2013). Cytokines as biomarkers of nanoparticle immunotoxicity. *Chemical Society Reviews, 42*(12), 5552–5576.

Engin, A. B., & Hayes, A. W. (2018). The impact of immunotoxicity in evaluation of the nanomaterials safety. *Toxicology Research and Application, 2*, 2397847318755579.

Erhirhie, E. O., Ihekwereme, C. P., & Ilodigwe, E. E. (2018). Advances in acute toxicity testing: Strengths, weaknesses and regulatory acceptance. *Interdisciplinary Toxicology, 11*(1), 5–12.

Ettrup, K., Kounina, A., Hansen, S. F., Meesters, J. A., Vea, E. B., & Laurent, A. (2017). Development of comparative toxicity potentials of TiO2 nanoparticles for use in life cycle assessment. *Environmental Science and Technology, 51*(7), 4027–4037.

Fadeel, B., Bussy, C., Merino, S., Vázquez, E., Flahaut, E., Mouchet, F., et al. (2018). Safety assessment of graphene-based materials: Focus on human health and the environment. *ACS Nano, 12*(11), 10582–10620.

Farnoud, A. M., & Nazemidashtarjandi, S. (2019). Emerging investigator series: Interactions of engineered nanomaterials with the cell plasma membrane; what have we learned from membrane models? *Environmental Science: Nano, 6*(1), 13–40.

Federico, A., Morgillo, F., Tuccillo, C., Ciardiello, F., & Loguercio, C. (2007). Chronic inflammation and oxidative stress in human carcinogenesis. *International Journal of Cancer, 121*(11), 2381–2386.

Ferdousi, R., Safdari, R., & Omidi, Y. (2017). Computational prediction of drug-drug interactions based on drugs functional similarities. *Journal of Biomedical Informatics, 70*, 54–64.

Fornaguera, C., & García-Celma, M. J. (2017). Personalized nanomedicine: A revolution at the nanoscale. *Journal of Personalized Medicine, 7*(4), 12.

Fu, P. P., Xia, Q., Hwang, H. M., Ray, P. C., & Yu, H. (2014). Mechanisms of nanotoxicity: Generation of reactive oxygen species. *Journal of Food and Drug Analysis, 22*(1), 64–75.

Ganguly, P., Breen, A., & Pillai, S. C. (2018). Toxicity of nanomaterials: Exposure, pathways, assessment, and recent advances. *ACS Biomaterials Science and Engineering, 4*(7), 2237–2275.

Gao, S., Yang, D., Fang, Y., Lin, X., Jin, X., Wang, Q., et al. (2019). Engineering nanoparticles for targeted remodeling of the tumor microenvironment to improve cancer immunotherapy. *Theranostics, 9*(1), 126.

Gerber, W., Steyn, J. D., Kotzé, A. F., & Hamman, J. H. (2018). Beneficial pharmacokinetic drug interactions: A tool to improve the bioavailability of poorly permeable drugs. *Pharmaceutics, 10*(3), 106.

Gilbertson, L. M., Zimmerman, J. B., Plata, D. L., Hutchison, J. E., & Anastas, P. T. (2015). Designing nanomaterials to maximize performance and minimize undesirable implications guided by the Principles of Green Chemistry. *Chemical Society Reviews, 44*(16), 5758–5777.

Gobbo, O. L., Sjaastad, K., Radomski, M. W., Volkov, Y., & Prina-Mello, A. (2015). Magnetic nanoparticles in cancer theranostics. *Theranostics, 5*(11), 1249.

Goodson, W. H., III, Lowe, L., Carpenter, D. O., Gilbertson, M., Manaf Ali, A., Lopez de Cerain Salsamendi, A., et al. (2015). Assessing the carcinogenic potential of low-dose exposures to chemical mixtures in the environment: the challenge ahead. *Carcinogenesis, 36*(Suppl_1), S254–S296.

Guo, Y. Y., Zhang, J., Zheng, Y. F., Yang, J., & Zhu, X. Q. (2011). Cytotoxic and genotoxic effects of multi-wall carbon nanotubes on human umbilical vein endothelial cells in vitro. *Mutation Research/Genetic Toxicology and Environmental Mutagenesis, 721*(2), 184–191.

Gupta, R. C., Chang, D., Nammi, S., Bensoussan, A., Bilinski, K., & Roufogalis, B. D. (2017). Interactions between antidiabetic drugs and herbs: An overview of mechanisms of action and clinical implications. *Diabetology and Metabolic Syndrome, 9*(1), 59.

Gurunathan, S., Kang, M. H., Qasim, M., & Kim, J. H. (2018). Nanoparticle-mediated combination therapy: Two-in-one approach for cancer. *International Journal of Molecular Sciences, 19*(10), 3264.

Halamoda-Kenzaoui, B., & Bremer-Hoffmann, S. (2018). Main trends of immune effects triggered by nanomedicines in preclinical studies. *International Journal of Nanomedicine, 13*, 5419.

Hardy, A., Benford, D., Halldorsson, T., Jeger, M. J., Knutsen, H. K., More, S., et al. (2018). Guidance on risk assessment of the application of nanoscience and nanotechnologies in the food and feed chain: Part 1, human and animal health. *EFSA Journal, 16*(7), 5327.

Hobson, D. W., Roberts, S. M., Shvedova, A. A., Warheit, D. B., Hinkley, G. K., & Guy, R. C. (2016). Applied nanotoxicology. *International Journal of Toxicology, 35*(1), 5–16.

Hua, S., De Matos, M. B., Metselaar, J. M., & Storm, G. (2018). Current trends and challenges in the clinical translation of nanoparticulate nanomedicines: Pathways for translational development and commercialization. *Frontiers in Pharmacology, 9*, 790.

Huang, Y. W., Cambre, M., & Lee, H. J. (2017). The toxicity of nanoparticles depends on multiple molecular and physicochemical mechanisms. *International Journal of Molecular Sciences, 18*(12), 2702.

Jeevanandam, J., Barhoum, A., Chan, Y. S., Dufresne, A., & Danquah, M. K. (2018). Review on nanoparticles and nanostructured materials: History, sources, toxicity and regulations. *Beilstein Journal of Nanotechnology, 9*(1), 1050–1074.

Jesus, S., Schmutz, M., Som, C., Borchard, G., Wick, P., & Borges, O. (2019). Hazard assessment of polymeric nanobiomaterials for drug delivery: What can we learn from literature so far. *Frontiers in Bioengineering and Biotechnology, 7*, 261.

Jiao, Q., Li, L., Mu, Q., & Zhang, Q. (2014). Immunomodulation of nanoparticles in nanomedicine applications. *BioMed Research International, 2014*, 426028.

Joshi, K., Mazumder, B., Chattopadhyay, P., Bora, N. S., Goyary, D., & Karmakar, S. (2019). Graphene family of nanomaterials: Reviewing advanced applications in drug delivery and medicine. *Current Drug Delivery, 16*(3), 195–214.

Jurj, A., Braicu, C., Pop, L. A., Tomuleasa, C., Gherman, C. D., & Berindan-Neagoe, I. (2017). The new era of nanotechnology, an alternative to change cancer treatment. *Drug Design, Development and Therapy, 11*, 2871.

Kashif, M., Andersson, C., Hassan, S., Karlsson, H., Senkowski, W., Fryknäs, M., et al. (2015). In vitro discovery of promising anti-cancer drug combinations using iterative maximisation of a therapeutic index. *Scientific Reports, 5*, 14118.

Kleandrova, V. V., Luan, F., Speck-Planche, A., & Cordeiro, N. D. (2015). In silico assessment of the acute toxicity of chemicals: Recent advances and new model for multitasking prediction of toxic effect. *Mini Reviews in Medicinal Chemistry, 15*(8), 677–686.

Kong, B., Seog, J. H., Graham, L. M., & Lee, S. B. (2011). Experimental considerations on the cytotoxicity of nanoparticles. *Nanomedicine, 6*(5), 929–941.

Kramer, J. A., Sagartz, J. E., & Morris, D. L. (2007). The application of discovery toxicology and pathology towards the design of safer pharmaceutical lead candidates. *Nature Reviews Drug Discovery, 6*(8), 636.

Krewski, D., Acosta, D., Jr., Andersen, M., Anderson, H., Bailar, J. C., III, Boekelheide, K., et al. (2010). Toxicity testing in the 21st century: A vision and a strategy. *Journal of Toxicology and Environmental Health, Part B, 13*(2–4), 51–138.

Kumar, A., Chen, F., Mozhi, A., Zhang, X., Zhao, Y., Xue, X., et al. (2013). Innovative pharmaceutical development based on unique properties of nanoscale delivery formulation. *Nanoscale, 5*(18), 8307–8325.

Kurutas, E. B. (2015). The importance of antioxidants which play the role in cellular response against oxidative/nitrosative stress: Current state. *Nutrition Journal, 15*(1), 71.

Kwon, J. Y., Koedrith, P., & Seo, Y. R. (2014). Current investigations into the genotoxicity of zinc oxide and silica nanoparticles in mammalian models in vitro and in vivo: Carcinogenic/genotoxic potential, relevant mechanisms and biomarkers, artifacts, and limitations. *International Journal of Nanomedicine, 9*(Suppl 2), 271.

La-Beck, N. M., Liu, X., & Wood, L. M. (2019). Harnessing liposome interactions with the immune system for the next breakthrough in cancer drug delivery. *Frontiers in Pharmacology, 10*, 220.

Laux, P., Tentschert, J., Riebeling, C., Braeuning, A., Creutzenberg, O., Epp, A., et al. (2018). Nanomaterials: Certain aspects of application, risk assessment and risk communication. *Archives of Toxicology, 92*(1), 121–141.

Lavik, E., & von Recum, H. (2011). The role of nanomaterials in translational medicine. *ACS Nano, 5*(5), 3419–3424.

Lee, D., Seo, Y., Khan, M. S., Hwang, J., Jo, Y., Son, J., et al. (2018). Use of nanoscale materials for the effective prevention and extermination of bacterial biofilms. *Biotechnology and Bioprocess Engineering, 23*(1), 1–10.

Li, Y., Ayala-Orozco, C., Rauta, P. R., & Krishnan, S. (2019). The application of nanotechnology in enhancing immunotherapy for cancer treatment: Current effects and perspective. *Nanoscale, 11*(37), 17157–17178.

Lim, S., Park, J., Shim, M. K., Um, W., Yoon, H. Y., Ryu, J. H., et al. (2019). Recent advances and challenges of repurposing nanoparticle-based drug delivery systems to enhance cancer immunotherapy. *Theranostics, 9*(25), 7906.

Lin, A., Giuliano, C. J., Palladino, A., John, K. M., Abramowicz, C., Yuan, M. L., et al. (2019). Off-target toxicity is a common mechanism of action of cancer drugs undergoing clinical trials. *Science Translational Medicine, 11*(509), eaaw8412.

Liu, Y., Peng, J., Wang, S., Xu, M., Gao, M., Xia, T., et al. (2018). Molybdenum disulfide/graphene oxide nanocomposites show favorable lung targeting and enhanced drug loading/tumor-killing efficacy with improved biocompatibility. *NPG Asia Materials, 10*(1), e458.

Lombardo, D., Kiselev, M. A., & Caccamo, M. T. (2019). Smart nanoparticles for drug delivery application: Development of versatile nanocarrier platforms in biotechnology and nanomedicine. *Journal of Nanomaterials, 2019*, 3702518.

Manke, A., Wang, L., & Rojanasakul, Y. (2013). Mechanisms of nanoparticle-induced oxidative stress and toxicity. *BioMed Research International, 2013*, 942916.

Marchant, G. E., Sylvester, D. J., & Abbott, K. W. (2008). Risk management principles for nanotechnology. *NanoEthics, 2*(1), 43–60.

McCallion, C., Burthem, J., Rees-Unwin, K., Golovanov, A., & Pluen, A. (2016). Graphene in therapeutics delivery: Problems, solutions and future opportunities. *European Journal of Pharmaceutics and Biopharmaceutics, 104*, 235–250.

Morales-Dalmau, J., Vilches, C., Sanz, V., de Miguel, I., Rodríguez-Fajardo, V., Berto, P., et al. (2019). Quantification of gold nanoparticle accumulation in tissue by two-photon luminescence microscopy. *Nanoscale, 11*(23), 11331.

Mourdikoudis, S., Pallares, R. M., & Thanh, N. T. (2018). Characterization techniques for nanoparticles: Comparison and complementarity upon studying nanoparticle properties. *Nanoscale, 10*(27), 12871–12934.

Mu, Q., Yu, J., McConnachie, L. A., Kraft, J. C., Gao, Y., Gulati, G. K., & Ho, R. J. (2018). Translation of combination nanodrugs into nanomedicines: Lessons learned and future outlook. *Journal of Drug Targeting, 26*(5–6), 435–447.

Müller, K., Bugnicourt, E., Latorre, M., Jorda, M., Echegoyen Sanz, Y., Lagaron, J. M., et al. (2017). Review on the processing and properties of polymer nanocomposites and nanocoatings and their applications in the packaging, automotive and solar energy fields. *Nanomaterials, 7*(4), 74.

Naskar, A., & Kim, K. S. (2019). Nanomaterials as delivery vehicles and components of new strategies to combat bacterial infections: Advantages and limitations. *Microorganisms, 7*(9), 356.

Navya, P. N., Kaphle, A., Srinivas, S. P., Bhargava, S. K., Rotello, V. M., & Daima, H. K. (2019). Current trends and challenges in cancer management and therapy using designer nanomaterials. *Nano Convergence, 6*(1), 23.

Neagu, M., Piperigkou, Z., Karamanou, K., Engin, A. B., Docea, A. O., Constantin, C., et al. (2017). Protein bio-corona: critical issue in immune nanotoxicology. *Archives of Toxicology, 91*(3), 1031–1048.

Nel, A., Xia, T., Meng, H., Wang, X., Lin, S., Ji, Z., & Zhang, H. (2012). Nanomaterial toxicity testing in the 21st century: Use of a predictive toxicological approach and high-throughput screening. *Accounts of Chemical Research, 46*(3), 607–621.

Nikalje, A. P. (2015). Nanotechnology and its applications in medicine. *Medicinal Chemistry, 5*(2), 081–089.

Oberdörster, G., Maynard, A., Donaldson, K., Castranova, V., Fitzpatrick, J., Ausman, K., et al. (2005). Principles for characterizing the potential human health effects from exposure to nanomaterials: Elements of a screening strategy. *Particle and Fibre Toxicology, 2*(1), 8.

Onoue, S., Yamada, S., & Chan, H. K. (2014). Nanodrugs: Pharmacokinetics and safety. *International Journal of Nanomedicine, 9*, 1025.

Paraskevaidi, M., Martin-Hirsch, P. L., Kyrgiou, M., & Martin, F. L. (2017). Underlying role of mitochondrial mutagenesis in the pathogenesis of a disease and current approaches for translational research. *Mutagenesis, 32*(3), 335–342.

Patra, J. K., Das, G., Fraceto, L. F., Campos, E. V. R., del Pilar Rodriguez-Torres, M., Acosta-Torres, L. S., et al. (2018). Nano based drug delivery systems: Recent developments and future prospects. *Journal of Nanobiotechnology, 16*(1), 71.

Prado-Audelo, D., María, L., Caballero-Florán, I. H., Meza-Toledo, J. A., Mendoza-Muñoz, N., González-Torres, M., et al. (2019). Formulations of curcumin nanoparticles for brain diseases. *Biomolecules, 9*(2), 56.

Qiu, T. A., Clement, P. L., & Haynes, C. L. (2018). Linking nanomaterial properties to biological outcomes: Analytical chemistry challenges in nanotoxicology for the next decade. *Chemical Communications, 54*(91), 12787–12803.

Radomska, A., Leszczyszyn, J., & Radomski, M. W. (2016). The nanopharmacology and nanotoxicology of nanomaterials: New opportunities and challenges. *Advances in Clinical and Experimental Medicine, 25*(1), 151–162.

Raies, A. B., & Bajic, V. B. (2016). In silico toxicology: Computational methods for the prediction of chemical toxicity. *Wiley Interdisciplinary Reviews: Computational Molecular Science, 6*(2), 147–172.

Rampado, R., Crotti, S., Caliceti, P., Pucciarelli, S., & Agostini, M. (2019). Nanovectors design for theranostic applications in colorectal cancer. *Journal of Oncology, 2019*, 2740923.

Ray, P. C., Yu, H., & Fu, P. P. (2009). Toxicity and environmental risks of nanomaterials: Challenges and future needs. *Journal of Environmental Science and Health Part C, 27*(1), 1–35.

Riediker, M., Zink, D., Kreyling, W., Oberdörster, G., Elder, A., Graham, U., et al. (2019). Particle toxicology and health-where are we? *Particle and Fibre Toxicology, 16*(1), 19.

Rizvi, S. A., & Saleh, A. M. (2018). Applications of nanoparticle systems in drug delivery technology. *Saudi Pharmaceutical Journal, 26*(1), 64–70.

Roberti, A., Valdes, A. F., Torrecillas, R., Fraga, M. F., & Fernandez, A. F. (2019). Epigenetics in cancer therapy and nanomedicine. *Clinical Epigenetics, 11*(1), 81.

Sahlgren, C., Meinander, A., Zhang, H., Cheng, F., Preis, M., Xu, C., et al. (2017). Tailored approaches in drug development and diagnostics: From molecular design to biological model systems. *Advanced Healthcare Materials, 6*(21), 1700258.

Sahu, D., Kannan, G. M., Tailang, M., & Vijayaraghavan, R. (2016). In vitro cytotoxicity of nanoparticles: A comparison between particle size and cell type. *Journal of Nanoscience, 2016*, 4023852.

Sang, W., Zhang, Z., Dai, Y., & Chen, X. (2019). Recent advances in nanomaterial-based synergistic combination cancer immunotherapy. *Chemical Society Reviews, 48*, 3771.

Schwarz-Plaschg, C., Kallhoff, A., & Eisenberger, I. (2017). Making nanomaterials safer by design? *NanoEthics, 3*(11), 277–281.

Seaton, A., Tran, L., Aitken, R., & Donaldson, K. (2009). Nanoparticles, human health hazard and regulation. *Journal of the Royal Society Interface, 7*(Suppl_1), S119–S129.

Senapati, V. A., Kumar, A., Gupta, G. S., Pandey, A. K., & Dhawan, A. (2015). ZnO nanoparticles induced inflammatory response and genotoxicity in human blood cells: A mechanistic approach. *Food and Chemical Toxicology, 85*, 61–70.

Senchukova, M. (2019). A brief review about the role of nanomaterials, mineral-organic nanoparticles, and extra-bone calcification in promoting carcinogenesis and tumor progression. *Biomedicine, 7*(3), 65.

Sharma, A., Madhunapantula, S. V., & Robertson, G. P. (2012). Toxicological considerations when creating nanoparticle-based drugs and drug delivery systems. *Expert Opinion on Drug Metabolism and Toxicology, 8*(1), 47–69.

Sharma, P., Jang, N. Y., Lee, J. W., Park, B. C., Kim, Y. K., & Cho, N. H. (2019). Application of ZnO-based nanocomposites for vaccines and cancer immunotherapy. *Pharmaceutics, 11*(10), 493.

Sierra, M. I., Valdés, A., Fernández, A. F., Torrecillas, R., & Fraga, M. F. (2016). The effect of exposure to nanoparticles and nanomaterials on the mammalian epigenome. *International Journal of Nanomedicine, 11*, 6297.

Simeonidis, S., Koutsilieri, S., Vozikis, A., Cooper, D. N., Mitropoulou, C., & Patrinos, G. P. (2019). Application of economic evaluation to assess feasibility for reimbursement of genomic testing as part of personalized medicine interventions. *Frontiers in Pharmacology, 10*, 830.

Singh, A. P., Biswas, A., Shukla, A., & Maiti, P. (2019). Targeted therapy in chronic diseases using nanomaterial-based drug delivery vehicles. *Signal Transduction and Targeted Therapy, 4*(1), 1–21.

Smolkova, B., Dusinska, M., & Gabelova, A. (2017). Nanomedicine and epigenome. Possible health risks. *Food and Chemical Toxicology, 109*, 780–796.

Sonali, M. K. V., Singh, R. P., Agrawal, P., Mehata, A. K., Datta Maroti Pawde, N., Sonkar, R., & Muthu, M. S. (2018). Nanotheranostics: Emerging strategies for early diagnosis and therapy of brain cancer. *Nano, 2*(1), 70.

Spivak, M. Y., Bubnov, R. V., Yemets, I. M., Lazarenko, L. M., Tymoshok, N. O., & Ulberg, Z. R. (2013). Gold nanoparticles-the theranostic challenge for PPPM: Nanocardiology application. *EPMA Journal, 4*(1), 18.

Sunderland, K. S., Yang, M., & Mao, C. (2017). Phage-enabled nanomedicine: From probes to therapeutics in precision medicine. *Angewandte Chemie International Edition, 56*(8), 1964–1992.

Tajbakhsh, J. (2011). DNA methylation topology: Potential of a chromatin landmark for epigenetic drug toxicology. *Epigenomics, 3*(6), 761–770.

Tamargo, J., Le Heuzey, J. Y., & Mabo, P. (2015). Narrow therapeutic index drugs: A clinical pharmacological consideration to flecainide. *European Journal of Clinical Pharmacology, 71*(5), 549–567.

Tan, B. L., Norhaizan, M. E., & Winnie-Pui-Pui Liew, H. S. (2018). Antioxidant and oxidative stress: A mutual interplay in age-related diseases. *Frontiers in Pharmacology, 9*, 1162.

Tardiff, R. G., & Rodricks, J. V. (Eds.). (2013). *Toxic substances and human risk: Principles of data interpretation*. Springer Science and Business Media. Berlin/Heidelberg, Germany.

Tavares, A. M., Louro, H., Antunes, S., Quarré, S., Simar, S., De Temmerman, P. J., et al. (2014). Genotoxicity evaluation of nanosized titanium dioxide, synthetic amorphous silica and multi-walled carbon nanotubes in human lymphocytes. *Toxicology In Vitro, 28*(1), 60–69.

Teow, Y., Asharani, P. V., Hande, M. P., & Valiyaveettil, S. (2011). Health impact and safety of engineered nanomaterials. *Chemical Communications, 47*(25), 7025–7038.

Thomson, J. P., Lempiäinen, H., Hackett, J. A., Nestor, C. E., Müller, A., Bolognani, F., et al. (2012). Non-genotoxic carcinogen exposure induces defined changes in the 5-hydroxymethylome. *Genome Biology, 13*(10), R93.

Trimble, W. S., & Grinstein, S. (2015). Barriers to the free diffusion of proteins and lipids in the plasma membrane. *The Journal of Cell Biology, 208*(3), 259–271.

Tsatsakis, A. M., Vassilopoulou, L., Kovatsi, L., Tsitsimpikou, C., Karamanou, M., Leon, G., et al. (2018). The dose response principle from philosophy to modern toxicology: The impact of ancient philosophy and medicine in modern toxicology science. *Toxicology Reports, 5*, 1107–1113.

Tuntland, T., Ethell, B., Kosaka, T., Blasco, F., Zang, R. X., Jain, M., et al. (2014). Implementation of pharmacokinetic and pharmacodynamic strategies in early research phases of drug discovery and development at Novartis Institute of Biomedical Research. *Frontiers in Pharmacology, 5*, 174.

ud Din, F., Aman, W., Ullah, I., Qureshi, O. S., Mustapha, O., Shafique, S., & Zeb, A. (2017). Effective use of nanocarriers as drug delivery systems for the treatment of selected tumors. *International Journal of Nanomedicine, 12*, 7291.

Ventola, C. L. (2017). Progress in nanomedicine: Approved and investigational nanodrugs. *Pharmacy and Therapeutics, 42*(12), 742.

Vimbela, G. V., Ngo, S. M., Fraze, C., Yang, L., & Stout, D. A. (2017). Antibacterial properties and toxicity from metallic nanomaterials. *International Journal of Nanomedicine, 12*, 3941.

Vitorino, C. V. (2018). Nanomedicine: Principles, properties and regulatory issues. *Frontiers in Chemistry, 6*, 360.

Warheit, D. B. (2018). Hazard and risk assessment strategies for nanoparticle exposures: How far have we come in the past 10 years? *F1000Research, 7*, 376.

Wen, H., Dan, M., Yang, Y., Lyu, J., Shao, A., Cheng, X., et al. (2017). Acute toxicity and genotoxicity of silver nanoparticle in rats. *PLoS One, 12*(9), e0185554.

Williams, D., Amman, M., Autrup, H., Bridges, J., Cassee, F., & Donaldson, K., et al. (2005). *The appropriateness of existing methodologies to assess the potential risks associated with engineered and adventitious products of nanotechnologies*. Report for the European Commission Health and Consumer Protection Directorate General by the Scientific Committee on Emerging and Newly Identified Health Risks, Brussels.

Wolfram, J., Zhu, M., Yang, Y., Shen, J., Gentile, E., Paolino, D., et al. (2015). Safety of nanoparticles in medicine. *Current Drug Targets, 16*(14), 1671–1681.

Xing, Y., Zhao, J., Conti, P. S., & Chen, K. (2014). Radiolabeled nanoparticles for multimodality tumor imaging. *Theranostics, 4*(3), 290.

Yang, N. J., & Hinner, M. J. (2015). Getting across the cell membrane: An overview for small molecules, peptides, and proteins. In *Site-specific protein labeling* (pp. 29–53). New York: Humana Press.

Yazdimamaghani, M., Moos, P. J., Dobrovolskaia, M. A., & Ghandehari, H. (2018). Genotoxicity of amorphous silica nanoparticles: Status and prospects. *Nanomedicine, 16*, 106.

Yu, Y., Zhang, Q., Mu, Q., Zhang, B., & Yan, B. (2008). Exploring the immunotoxicity of carbon nanotubes. *Nanoscale Research Letters, 3*(8), 271.

Yu, D., Kahen, E., Cubitt, C. L., McGuire, J., Kreahling, J., Lee, J., et al. (2015). Identification of synergistic, clinically achievable, combination therapies for osteosarcoma. *Scientific Reports, 5*, 16991.

Yuan, Y. G., Zhang, S., Hwang, J. Y., & Kong, I. K. (2018). Silver nanoparticles potentiates cytotoxicity and apoptotic potential of camptothecin in human cervical cancer cells. *Oxidative Medicine and Cellular Longevity, 2018*, 6121328.

Yuan, X., Zhang, X., Sun, L., Wei, Y., & Wei, X. (2019). Cellular toxicity and immunological effects of carbon-based nanomaterials. *Particle and Fibre Toxicology, 16*(1), 18.

Zhang, X. Q., Xu, X., Bertrand, N., Pridgen, E., Swami, A., & Farokhzad, O. C. (2012). Interactions of nanomaterials and biological systems: Implications to personalized nanomedicine. *Advanced Drug Delivery Reviews, 64*(13), 1363–1384.

Zottel, A., Videtič Paska, A., & Jovčevska, I. (2019). Nanotechnology meets oncology: Nanomaterials in brain cancer research, diagnosis and therapy. *Materials, 12*(10), 1588.

Chapter 5
ADMETox: Bringing Nanotechnology Closer to Lipinski's Rule of Five

As a man who has devoted his whole life to the most clearheaded science, to the study of matter, I can tell you as a result of my research about the atoms this much: There is no matter as such! All matter originates and exists only by virtue of a force which brings the particles of an atom to vibration and holds this most minute solar system of the atom together. . . . We must assume behind this force the existence of a conscious and intelligent Mind. This Mind is the matrix of all matter
New scientific ideas never spring from a communal body, however organized, but rather from the head of an individually inspired researcher who struggles with his problems in lonely thought and unites all his thought on one single point which is his whole world for the moment
— Max Planck (1858–1947)

Abstract To develop medications with a good therapeutic index, it is necessary to analyze the concepts of contents in Lipinski's rules, as well as the druglikeness parameters that make up the ADME (absorption, distribution, metabolism, and excretion) score. These parameters consider important factors of the behavior of a drug within the human body in order to favor a good therapy with great pharmacological potential and low toxicity. Thus, in this order of ideas, the development of modern antimicrobial nanopharmaceuticals requires that all possible factors that alter the maintenance of anti-infective inhibitory concentrations within the body be controlled, which will allow an effective cure and prevent the emergence of resistance. For this reason, the objective of this chapter is the analysis of the parameters that are part of the ADME score in light of the use of nanomaterials for the design and development of new medications, which will allow obtaining medications with greater pharmacological activity and with a high rate of security.

J. Bueno, *Preclinical Evaluation of Antimicrobial Nanodrugs*, Nanotechnology in the Life Sciences, https://doi.org/10.1007/978-3-030-43855-5_5

5.1 Introduction

The search for druglikeness requires establishing the chemical, physical, and biological properties that a medicine needs to be successful when administered to the sick patient (Hughes et al. 2011; Mohs and Greig 2017). Likewise, the biological properties of druglikeness should be correlated with molecular ones, as well as with pharmacokinetics in order to develop highly active drugs with less adverse effects (Zhang et al. 2015; Daina et al. 2017; Machado et al. 2018; Olğaç et al. 2019). Thus, the search for druglikeness requires the correlation of ADMEtox in which the absorption, distribution, metabolism, and excretion of a molecule is evaluated based on the toxicity it presents during its exposure (Fig. 5.1) (Tsaioun et al. 2016; Dong et al. 2018; Guan et al. 2019; Han et al. 2019). Likewise, the ADMET score should include the solubility of the nano-medication, the potency of the action on the therapeutic target, lipophilicity, the efficiency as a ligand, and its molecular weight to predict its diffusion and absorption (Charifson and Walters 2014; ud Din et al. 2017). In this order of ideas, adapting Lipinski's rule of five establishes the score that determines how much capacity a molecule or formulation has in becoming a drug, which determine a correlation between pharmacological activity, lipophilicity, and absorption (Van De Waterbeemd et al. 2001; Liang et al. 2008; Gunasekaran et al. 2014; Meanwell 2016). Thus, in a nanotechnological approach, the ADMET score is determined by the biological activity that the physical properties of nanomaterials exert within a chemical structure, which gives them a pharmacological action (Wong et al. 2013; Tian et al. 2015; Navya and Daima 2016; Ulbrich et al. 2016). Likewise, in this translation from physics to pharmacology, the biological effects of chemical compounds are enhanced by changing their interaction and molecular signaling properties, so it is necessary to assume the nanocomposites as a

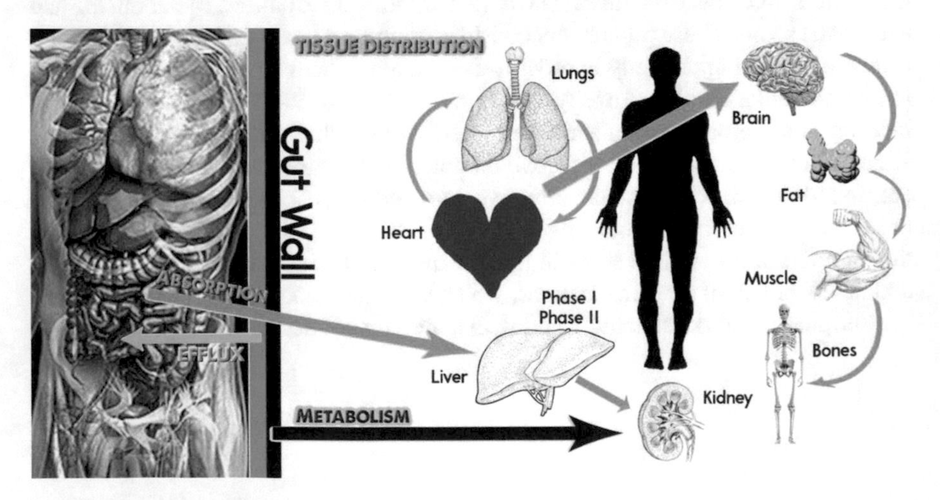

Fig. 5.1 ADMETox parameters

spatial reconfiguration of medicines, and this requires complementary rules of druglikeness (Schenone et al. 2013; Tubert-Brohman et al. 2013; Scheinberg et al. 2017). Based on the above, the objective of this chapter is to analyze how in nanomedicine the biological activity of drugs is influenced by the physical properties of nanomaterials to establish a new adaptation of Lipinski's rules that will allow the development of new drug discovery programs (Patra et al. 2018; Sharma and Hussain 2018; Vitorino 2018).

5.2 ADMET Score

Biopharmaceutical Classification System (BCS) classifies compounds class IV as those that have low solubility and low permeability; these types of molecules attract the attention of nanomedicine as a group of compounds to be impacted on their therapeutic index (Papich and Martinez 2015; Emami et al. 2018; Kumar et al. 2018; Boyd et al. 2019). Thus, in order to improve the therapeutic index and increase activity by decreasing toxicity, it is necessary to evaluate the parameters of the ADME profile based on lipophilicity, molecular weight, and chemical structure (Di and Kerns 2015; Mignani et al. 2018; Johnson et al. 2018; Lucas et al. 2019). This ADME profile in nanomaterials must be correlated with other aspects such as the induction of reactive oxygen species (ROS) that is one of the main mechanisms that mediate nanotoxicity (Manke et al. 2013; Fu et al. 2014; Abdal Dayem et al. 2017; Brohi et al. 2017). Thus, in compliance with Lipinski's rule of five in nano-toxicology, it is a priority to determine the interaction between the induction of ROS and lipophilicity, as well as the affectation of a molecular structure that can have more than five donors and more than ten hydrogen bond acceptors (Fig. 5.2) (Khanna and Ranganathan 2009; Benet et al. 2016; Bocci et al. 2019; Su et al. 2019). Likewise, these interactions of nanomaterials in their toxicity must be related to each parameter of druglikeness in order to prevent possible problems with the designed nanomedicaments, especially with the first factor that is absorption (De Jong and Borm 2008; Buse and El-Aneed 2010; Lee et al. 2015; Rizvi and Saleh 2018). In this way the absorption establishes the route of entry of the nanomaterials, as well as their ability to cross the biological membranes which determines the initial step of the interaction of the nanocomposites with the biological processes of

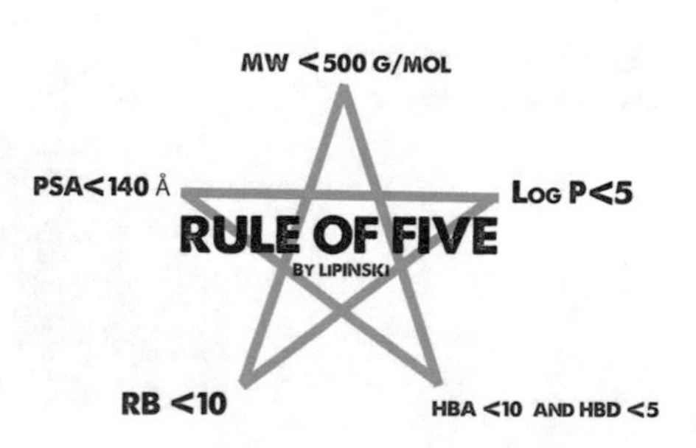

Fig. 5.2 Lipinski's rule of five

the patient, so it is considered of radical importance in the discovery of new drugs (Zhang et al. 2012; Baptista et al. 2018; Nekoueian et al. 2019).

5.3 Absorption and Interaction with Biological Systems in Nanotechnology

The mechanism of absorption of nanomaterials is linked to their ability to spread and cross the biological barriers due to lipophilicity that they present (Fig. 5.3) (Upadhyay 2014; Gnach et al. 2015; Zhou et al. 2018; Brandelli 2019). This ability of nanocomposites to cross biological barriers can favor tissue damage in absorption routes by inducing cytotoxicity and irritation (Sharma et al. 2012; Banerjee 2018; Chenthamara et al. 2019; Choudhury et al. 2019). The routes of entry of nanomedicines that can be used are oral, inhalation, dermal, and parenteral, each with specific tissues and barriers, as well as with a different absorption rate, which determines the possibilities of application of each drug designed (Hamidi et al. 2013; Zhang et al. 2014; Ahmed and Aljaeid 2016; Chenthamara et al. 2019). Thus, there are methods that can be used to determine the absorption rate through the administration routes, such as the use of cell culture monolayers for in vitro permeability assays models such as Caco-2, which can be correlated with the cytotoxicity to obtain a reliable prediction of druglikeness (Sarmento et al. 2012; Larregieu and Benet 2013; Awortwe et al. 2014; Stockdale et al. 2019). In that order of ideas, nanoparticles have shown cellular damage in these types of cells, as well as genotoxicity, which could serve as a predictive factor of gastric intolerance of nanoformulations (He et al. 2013; Gossmann et al. 2018; Silva et al. 2019; Colombo et al. 2019). It has also been shown that in the process of transcytosis of nanomaterials through biological barriers, it releases proinflammatory cytokines, which may cause delayed irritation and hypersensitivity phenomena that would cause decreased absorption (Meng et al. 2018; Teleanu et al. 2018). Likewise, this activation of the immune system caused by nanomaterials in the routes of administration in the human body can lead to antigenic recognition by the presenting cells and T lymphocytes in order to induce autoimmune responses and allergic diseases (Zolnik et al. 2010; Dacoba et al. 2017; Dobrovolskaia 2019; Roach et al. 2019). On the other hand, the interaction of nanocomposites with host proteins that may result in

Fig. 5.3 Absorption of nanomaterials

immunocomplex accumulation as well as toxicity in tissues such as renal and hepatic at the time of nanodrug distribution after absorption must be taken into account (Nehoff et al. 2014; Blanco et al. 2015; Pelaz et al. 2017).

5.4 Distribution: Specificity After Absorption

One of the major limitations of nanomedicaments is their stable binding to proteins to be distributed after absorption, as well as specificity on the target organ, which decreases their accumulation and toxicity (Singh and Lillard 2009; Howard et al. 2014; Auría-Soro et al. 2019; Bailly et al. 2019). In this order of ideas, the inappropriate distribution of a nanocomposite prevents reaching adequate therapeutic concentrations that allow to reach infected tissues and thus produce tissue damage when excreted (Pattnaik et al. 2016; Kermanizadeh et al. 2018; Qiao et al. 2019; Singh et al. 2019). Likewise, the possibility of accumulating metabolites of nanopharmaceutical degradation products increases with the decrease in distribution, as well as the decrease in pharmacological activity, which causes therapeutic failures (Uchegbu et al. 2013; Hare et al. 2017; Bogart et al. 2014; Zhang et al. 2019). To ensure that nanomaterials have a correct distribution after absorption, they can be associated with nanostructures with specific ligands of both plasma proteins and the target organ that will allow therapeutic levels to be maintained and nanocomposites not to accumulate in other tissues (Conde et al. 2014; Shukla et al. 2016; Clegg et al. 2019; Li et al. 2019). Thus, the nanomaterials associated with cell ligands have become a promising development strategy that together with fluorescent markers will allow to determine the concentrations of the nanopharmaceutical, as well as its accumulation in tissues with adequate therapeutic levels (Jain 2010; Mirkin et al. 2011; Bejarano et al. 2018). Likewise, the specificity in the distribution of the medicines allows to initiate an adequate metabolism of the same when they have already carried out their pharmacological activity preventing their subsequent accumulation in the host tissues (Oberdörster et al. 2005; Behzadi et al. 2017; Neagu et al. 2017; Fadeel et al. 2018). In the same way, guaranteeing the adequate distribution of the nanomaterial will allow the metabolism from the tissues to the excretion in the liver and kidney to be more efficient and prevent the toxic consequences of accumulation (Longmire et al. 2008; Pizzorno 2015; Recordati et al. 2015; Volkov et al. 2017).

5.5 Metabolism: Nanomaterial Clearance

The nanomaterials are cleared by Kupffer cells, B lymphocytes, and liver cells, which is why toxicity in this type of cells should be previously established as part of the design and development of nanopharmaceuticals (Watson et al. 2015; Tsoi et al. 2016; Tavares et al. 2017). Likewise, the liver and spleen become the major biological barriers of nanomedicines due to their ability to sequester nanocomposites from the bloodstream preventing their distribution (Almeida et al. 2011; Bhattacharya et al. 2016; Park et al. 2016; Mohammadi et al. 2017). In this order of ideas, the nanoparticles are metabolized through the hepatic beta-glucuronidase pathway to be excreted through the bile so that all metabolism should be determined from the clearance and liver function (Chadha et al. 2015; Ge et al. 2016; Kitchin et al. 2017). Thus the liver function and the activity of nanomaterials on liver enzymes should be determined and analyzed in order to establish the safety and avoid the hepatotoxicity risks that nanomedicines represent (Henkler et al. 2012; Boverhof et al. 2015; Rauscher et al. 2017). Likewise, in order to obtain more biocompatible nanopharmaceuticals, it is important to examine the hepatocarcinogenic potential of each substance to be used (Curtis et al. 2006; Vega-Villa et al. 2008; Arora et al. 2012). Finally, the biliary excretion of nanomaterials determines a new challenge for the development of nanodrugs, because it limits their application in patients with liver failure and gastrointestinal disorder; liver clearance will also become the ideal biomarker to predict and avoid nanotoxicity (Zhang et al. 2016; Pietroiusti et al. 2017; Zhu et al. 2019).

5.6 Nanopharmaceutical Excretion: The Risk of Gastrointestinal Disorders

After glucuronidation the excretion of nanomaterials is performed through bile and gastrointestinal tract through feces; this opens the possibility of toxicity of excreted metabolites on the gastrointestinal system, as well as the appearance of dysfunctions and carcinogenicity (Bergin and Witzmann 2013). Likewise, the role of the intestinal microbiota in nanotoxicology becomes an issue to be considered in order to improve the excretion mechanisms and boost liver clearance (Heringa et al. 2016; Mao et al. 2016; Pietroiusti et al. 2016; Federico et al. 2017). In this order of ideas, the biotransformations that the gut microbiome can perform on excreted nanocomposites can induce new substances with potential risk and toxicity (Claus et al. 2016; Koppel et al. 2017; Wilson and Nicholson 2017; Guthrie et al. 2019). Thus the influence of nanomaterials on the index of *Firmicutes* and *Bacteroidetes* of the human microbiome is an important factor to consider in order to avoid possible disorders and limitations in excretion (Kho and Lal 2018; Chen et al. 2019; Rinninella et al. 2019). So in this way, the correct excretion of nanopharmaceuticals, as well as the knowledge of their intestinal biotransformation, will allow the development of novel strategies to avoid the accumulation and toxicity in tissues such as the use of hepatoprotective agents such as quercetin to allow liver metabolism (Eftekhari et al. 2018).

5.7 Conclusions

Thus, the correct establishment of ADME score parameters is an important tool to increase bioavailability, improve pharmacological action, and avoid toxicity (Daina et al. 2017; Guan et al. 2019). In this order of ideas, the monitoring of the ADME score in the design of nanomedicaments will allow to increase its druglikeness and the success of the therapies implemented for the development of therapeutic alternatives (Zhang and Wilkinson 2007; Csermely et al. 2013). Likewise, Lipinski's rules should incorporate the concepts of human microbiome and biotransformation in order to obtain drugs with a better therapeutic index (Sonnenburg and Fischbach 2011; Sharma et al. 2017).

Acknowledgments The author thanks Sebastian Ritoré for his collaboration and invaluable support during the writing of this chapter, as well as the graphics contained in this book.

References

Abdal Dayem, A., Hossain, M. K., Lee, S. B., Kim, K., Saha, S. K., Yang, G. M., et al. (2017). The role of reactive oxygen species (ROS) in the biological activities of metallic nanoparticles. *International Journal of Molecular Sciences, 18*(1), 120.

Ahmed, T. A., & Aljaeid, B. M. (2016). Preparation, characterization, and potential application of chitosan, chitosan derivatives, and chitosan metal nanoparticles in pharmaceutical drug delivery. *Drug Design, Development and Therapy, 10*, 483.

Almeida, J. P. M., Chen, A. L., Foster, A., & Drezek, R. (2011). In vivo biodistribution of nanoparticles. *Nanomedicine, 6*(5), 815–835.

Arora, S., Rajwade, J. M., & Paknikar, K. M. (2012). Nanotoxicology and in vitro studies: The need of the hour. *Toxicology and Applied Pharmacology, 258*(2), 151–165.

Auría-Soro, C., Nesma, T., Juanes-Velasco, P., Landeira-Viñuela, A., Fidalgo-Gomez, H., Acebes-Fernandez, V., et al. (2019). Interactions of nanoparticles and biosystems: Microenvironment of nanoparticles and biomolecules in nanomedicine. *Nanomaterials, 9*(10), 1365.

Awortwe, C., Fasinu, P. S., & Rosenkranz, B. (2014). Application of Caco-2 cell line in herb-drug interaction studies: Current approaches and challenges. *Journal of Pharmacy and Pharmaceutical Sciences, 17*(1), 1.

Bailly, A. L., Correard, F., Popov, A., Tselikov, G., Chaspoul, F., Appay, R., et al. (2019). In vivo evaluation of safety, biodistribution and pharmacokinetics of laser-synthesized gold nanoparticles. *Scientific Reports, 9*(1), 1–12.

Banerjee, A. N. (2018). Graphene and its derivatives as biomedical materials: Future prospects and challenges. *Interface Focus, 8*(3), 20170056.

Baptista, P. V., McCusker, M. P., Carvalho, A., Ferreira, D. A., Mohan, N. M., Martins, M., & Fernandes, A. R. (2018). Nano-strategies to fight multidrug resistant bacteria—"A Battle of the Titans". *Frontiers in Microbiology, 9*, 1441.

Behzadi, S., Serpooshan, V., Tao, W., Hamaly, M. A., Alkawareek, M. Y., Dreaden, E. C., et al. (2017). Cellular uptake of nanoparticles: Journey inside the cell. *Chemical Society Reviews, 46*(14), 4218–4244.

Bejarano, J., Navarro-Marquez, M., Morales-Zavala, F., Morales, J. O., Garcia-Carvajal, I., Araya-Fuentes, E., et al. (2018). Nanoparticles for diagnosis and therapy of atherosclerosis and myocardial infarction: Evolution toward prospective theranostic approaches. *Theranostics, 8*(17), 4710.

Benet, L. Z., Hosey, C. M., Ursu, O., & Oprea, T. I. (2016). BDDCS, the rule of 5 and drugability. *Advanced Drug Delivery Reviews, 101*, 89–98.

Bergin, I. L., & Witzmann, F. A. (2013). Nanoparticle toxicity by the gastrointestinal route: Evidence and knowledge gaps. *International Journal of Biomedical Nanoscience and Nanotechnology, 3*(1–2).

Bhattacharya, K., Mukherjee, S. P., Gallud, A., Burkert, S. C., Bistarelli, S., Bellucci, S., et al. (2016). Biological interactions of carbon-based nanomaterials: From coronation to degradation. *Nanomedicine: Nanotechnology, Biology and Medicine, 12*(2), 333–351.

Blanco, E., Shen, H., & Ferrari, M. (2015). Principles of nanoparticle design for overcoming biological barriers to drug delivery. *Nature Biotechnology, 33*(9), 941.

Bocci, G., Benet, L. Z., & Oprea, T. I. (2019). Can BDDCS illuminate targets in drug design? *Drug Discovery Today, 24*(12), 2299.

Bogart, L. K., Pourroy, G., Murphy, C. J., Puntes, V., Pellegrino, T., Rosenblum, D., et al. (2014). Nanoparticles for imaging, sensing, and therapeutic intervention. *ACS Nano, 8*(4), 3107.

Boverhof, D. R., Bramante, C. M., Butala, J. H., Clancy, S. F., Lafranconi, M., West, J., & Gordon, S. C. (2015). Comparative assessment of nanomaterial definitions and safety evaluation considerations. *Regulatory Toxicology and Pharmacology, 73*(1), 137–150.

Boyd, B. J., Bergström, C. A., Vinarov, Z., Kuentz, M., Brouwers, J., Augustijns, P., et al. (2019). Successful oral delivery of poorly water-soluble drugs both depends on the intraluminal behavior of drugs and of appropriate advanced drug delivery systems. *European Journal of Pharmaceutical Sciences, 137*, 104967.

Brandelli, A. (2019). The interaction of nanostructured antimicrobials with biological systems: Cellular uptake, trafficking and potential toxicity. *Food Science and Human Wellness*. https://doi.org/10.1016/j.fshw.2019.12.003.

Brohi, R. D., Wang, L., Talpur, H. S., Wu, D., Khan, F. A., Bhattarai, D., et al. (2017). Toxicity of nanoparticles on the reproductive system in animal models: A review. *Frontiers in Pharmacology, 8*, 606.

Buse, J., & El-Aneed, A. (2010). Properties, engineering and applications of lipid-based nanoparticle drug-delivery systems: Current research and advances. *Nanomedicine, 5*(8), 1237–1260.

Chadha, N., Chaturvedi, S., Lal, S., Mishra, A. K., Pulicharla, R., Cledon, M., et al. (2015). Engineered nanoparticles associated metabolomics. *Journal of Hazardous, Toxic, and Radioactive Waste, 20*(1), B4015003.

Charifson, P. S., & Walters, W. P. (2014). Acidic and basic drugs in medicinal chemistry: A perspective. *Journal of Medicinal Chemistry, 57*(23), 9701–9717.

Chen, Z., Zhou, D., Han, S., Zhou, S., & Jia, G. (2019). Hepatotoxicity and the role of the gut-liver axis in rats after oral administration of titanium dioxide nanoparticles. *Particle and Fibre Toxicology, 16*(1), 1–17.

Chenthamara, D., Subramaniam, S., Ramakrishnan, S. G., Krishnaswamy, S., Essa, M. M., Lin, F. H., & Qoronfleh, M. W. (2019). Therapeutic efficacy of nanoparticles and routes of administration. *Biomaterials Research, 23*(1), 1–29.

Choudhury, H., Gorain, B., Pandey, M., Kaur, R., & Kesharwani, P. (2019). Strategizing biodegradable polymeric nanoparticles to cross the biological barriers for cancer targeting. *International Journal of Pharmaceutics, 565*, 509.

Claus, S. P., Guillou, H., & Ellero-Simatos, S. (2016). The gut microbiota: A major player in the toxicity of environmental pollutants? *NPJ Biofilms and Microbiomes, 2*, 16003.

Clegg, J. R., Irani, A. S., Ander, E. W., Ludolph, C. M., Venkataraman, A. K., Zhong, J. X., & Peppas, N. A. (2019). Synthetic networks with tunable responsiveness, biodegradation, and molecular recognition for precision medicine applications. *Science Advances, 5*(9), eaax7946.

Colombo, G., Cortinovis, C., Moschini, E., Bellitto, N., Perego, M. C., Albonico, M., et al. (2019). Cytotoxic and proinflammatory responses induced by ZnO nanoparticles in in vitro intestinal barrier. *Journal of Applied Toxicology, 39*, 1155.

Conde, J., Dias, J. T., Grazú, V., Moros, M., Baptista, P. V., & de la Fuente, J. M. (2014). Revisiting 30 years of biofunctionalization and surface chemistry of inorganic nanoparticles for nanomedicine. *Frontiers in Chemistry, 2*, 48.

Csermely, P., Korcsmáros, T., Kiss, H. J., London, G., & Nussinov, R. (2013). Structure and dynamics of molecular networks: A novel paradigm of drug discovery: A comprehensive review. *Pharmacology and Therapeutics, 138*(3), 333–408.

Curtis, J., Greenberg, M., Kester, J., Phillips, S., & Krieger, G. (2006). Nanotechnology and nanotoxicology. *Toxicological Reviews, 25*(4), 245–260.

Dacoba, T. G., Olivera, A., Torres, D., Crecente-Campo, J., & Alonso, M. J. (2017). Modulating the immune system through nanotechnology. *Seminars in Immunology, 34*, 78–102. Academic Press.

Daina, A., Michielin, O., & Zoete, V. (2017). SwissADME: A free web tool to evaluate pharmacokinetics, drug-likeness and medicinal chemistry friendliness of small molecules. *Scientific Reports, 7*, 42717.

De Jong, W. H., & Borm, P. J. (2008). Drug delivery and nanoparticles: Applications and hazards. *International Journal of Nanomedicine, 3*(2), 133.

Di, L., & Kerns, E. H. (2015). *Drug-like properties: Concepts, structure design and methods from ADME to toxicity optimization*. Academic. Academic Press, Cambridge, Massachusetts

Dobrovolskaia, M. A. (2019). Nucleic acid nanoparticles at a crossroads of vaccines and immunotherapies. *Molecules, 24*(24), 4620.

Dong, J., Wang, N. N., Yao, Z. J., Zhang, L., Cheng, Y., Ouyang, D., et al. (2018). ADMETlab: A platform for systematic ADMET evaluation based on a comprehensively collected ADMET database. *Journal of Cheminformatics, 10*(1), 29.

Eftekhari, A., Ahmadian, E., Panahi-Azar, V., Hosseini, H., Tabibiazar, M., & Maleki Dizaj, S. (2018). Hepatoprotective and free radical scavenging actions of quercetin nanoparticles on aflatoxin B1-induced liver damage: In vitro/in vivo studies. *Artificial Cells, Nanomedicine, and Biotechnology, 46*(2), 411–420.

Emami, S., Siahi-Shadbad, M., Adibkia, K., & Barzegar-Jalali, M. (2018). Recent advances in improving oral drug bioavailability by cocrystals. *BioImpacts: BI, 8*(4), 305.

Fadeel, B., Bussy, C., Merino, S., Vázquez, E., Flahaut, E., Mouchet, F., et al. (2018). Safety assessment of graphene-based materials: Focus on human health and the environment. *ACS Nano, 12*(11), 10582–10620.

Federico, A., Dallio, M., Caprio, G. G., Ormando, V. M., & Loguercio, C. (2017). Gut microbiota and the liver. *Minerva Gastroenterologica e Dietologica, 63*(4), 385–398.

Fu, P. P., Xia, Q., Hwang, H. M., Ray, P. C., & Yu, H. (2014). Mechanisms of nanotoxicity: Generation of reactive oxygen species. *Journal of Food and Drug Analysis, 22*(1), 64–75.

Ge, S., Tu, Y., & Hu, M. (2016). Challenges and opportunities with predicting in vivo phase II metabolism via glucuronidation from in vitro data. *Current Pharmacology Reports, 2*(6), 326–338.

Gnach, A., Lipinski, T., Bednarkiewicz, A., Rybka, J., & Capobianco, J. A. (2015). Upconverting nanoparticles: Assessing the toxicity. *Chemical Society Reviews, 44*(6), 1561–1584.

Gossmann, R., Spek, S., Langer, K., & Mulac, D. (2018). Didodecyldimethylammonium bromide (DMAB) stabilized poly (lactic-co-glycolic acid) (PLGA) nanoparticles: Uptake and cytotoxic potential in Caco-2 cells. *Journal of Drug Delivery Science and Technology, 43*, 430–438.

Guan, L., Yang, H., Cai, Y., Sun, L., Di, P., Li, W., et al. (2019). ADMET-score–a comprehensive scoring function for evaluation of chemical drug-likeness. *Medchemcomm, 10*(1), 148–157.

Gunasekaran, T., Haile, T., Nigusse, T., & Dhanaraju, M. D. (2014). Nanotechnology: An effective tool for enhancing bioavailability and bioactivity of phytomedicine. *Asian Pacific Journal of Tropical Biomedicine, 4*, S1–S7.

Guthrie, L., Wolfson, S., & Kelly, L. (2019). The human gut chemical landscape predicts microbe-mediated biotransformation of foods and drugs. *eLife, 8*, e42866.

Hamidi, M., Azadi, A., Rafiei, P., & Ashrafi, H. (2013). A pharmacokinetic overview of nanotechnology-based drug delivery systems: An ADME-oriented approach. *Critical Reviews™ in Therapeutic Drug Carrier Systems, 30*(5), 435.

Han, Y., Zhang, X., Zhang, J., & Hu, C. Q. (2019). In silico ADME and toxicity prediction of ceftazidime and its impurities. *Frontiers in Pharmacology, 10*, 434.

Hare, J. I., Lammers, T., Ashford, M. B., Puri, S., Storm, G., & Barry, S. T. (2017). Challenges and strategies in anti-cancer nanomedicine development: An industry perspective. *Advanced Drug Delivery Reviews, 108*, 25–38.

He, B., Lin, P., Jia, Z., Du, W., Qu, W., Yuan, L., et al. (2013). The transport mechanisms of polymer nanoparticles in Caco-2 epithelial cells. *Biomaterials, 34*(25), 6082–6098.

Henkler, F., Tralau, T., Tentschert, J., Kneuer, C., Haase, A., Platzek, T., et al. (2012). Risk assessment of nanomaterials in cosmetics: A European union perspective. *Archives of Toxicology, 86*(11), 1641–1646.

Heringa, M. B., Geraets, L., van Eijkeren, J. C., Vandebriel, R. J., de Jong, W. H., & Oomen, A. G. (2016). Risk assessment of titanium dioxide nanoparticles via oral exposure, including toxicokinetic considerations. *Nanotoxicology, 10*(10), 1515–1525.

Howard, M., Zern, B. J., Anselmo, A. C., Shuvaev, V. V., Mitragotri, S., & Muzykantov, V. (2014). Vascular targeting of nanocarriers: Perplexing aspects of the seemingly straightforward paradigm. *ACS Nano, 8*(5), 4100–4132.

Hughes, J. P., Rees, S., Kalindjian, S. B., & Philpott, K. L. (2011). Principles of early drug discovery. *British Journal of Pharmacology, 162*(6), 1239–1249.

Jain, K. K. (2010). Advances in the field of nanooncology. *BMC Medicine, 8*(1), 83.

Johnson, T. W., Gallego, R. A., & Edwards, M. P. (2018). Lipophilic efficiency as an important metric in drug design. *Journal of Medicinal Chemistry, 61*(15), 6401–6420.

Kermanizadeh, A., Powell, L. G., Stone, V., & Møller, P. (2018). Nanodelivery systems and stabilized solid-drug nanoparticles for orally administered medicine: Current landscape. *International Journal of Nanomedicine, 13*, 7575.

Khanna, V., & Ranganathan, S. (2009). Physicochemical property space distribution among human metabolites, drugs and toxins. *BMC Bioinformatics, 10*(15), S10. BioMed central.

Kho, Z. Y., & Lal, S. K. (2018). The human gut microbiome–a potential controller of wellness and disease. *Frontiers in Microbiology, 9*, 1835.

Kitchin, K. T., Stirdivant, S., Robinette, B. L., Castellon, B. T., & Liang, X. (2017). Metabolomic effects of CeO 2, SiO 2 and CuO metal oxide nanomaterials on HepG2 cells. *Particle and Fibre Toxicology, 14*(1), 50.

Koppel, N., Rekdal, V. M., & Balskus, E. P. (2017). Chemical transformation of xenobiotics by the human gut microbiota. *Science, 356*(6344), eaag2770.

Kumar, S., Kaur, R., Rajput, R., & Singh, M. (2018). Bio Pharmaceutics Classification System (BCS) class IV drug nanoparticles: Quantum leap to improve their therapeutic index. *Advanced Pharmaceutical Bulletin, 8*(4), 617.

Larregieu, C. A., & Benet, L. Z. (2013). Drug discovery and regulatory considerations for improving in silico and in vitro predictions that use Caco-2 as a surrogate for human intestinal permeability measurements. *The AAPS Journal, 15*(2), 483–497.

Lee, B. K., Yun, Y. H., & Park, K. (2015). Smart nanoparticles for drug delivery: Boundaries and opportunities. *Chemical Engineering Science, 125*, 158–164.

Li, C., Wang, J., Wang, Y., Gao, H., Wei, G., Huang, Y., et al. (2019). Recent progress in drug delivery. *Acta Pharmaceutica Sinica B, 9*(6), 1145.

Liang, X. J., Chen, C., Zhao, Y., Jia, L., & Wang, P. C. (2008). Biopharmaceutics and therapeutic potential of engineered nanomaterials. *Current Drug Metabolism, 9*(8), 697–709.

Longmire, M., Choyke, P. L., & Kobayashi, H. (2008). Clearance properties of nano-sized particles and molecules as imaging agents: Considerations and caveats. *Nanomedicine (London, England), 3*(5), 703.

Lucas, A. J., Sproston, J. L., Barton, P., & Riley, R. J. (2019). Estimating human ADME properties, pharmacokinetic parameters and likely clinical dose in drug discovery. *Expert Opinion on Drug Discovery, 14*(12), 1313–1327.

Machado, D., Girardini, M., Viveiros, M., & Pieroni, M. (2018). Challenging the drug-likeness dogma for new drug discovery in tuberculosis. *Frontiers in Microbiology, 9*, 1367.

Manke, A., Wang, L., & Rojanasakul, Y. (2013). Mechanisms of nanoparticle-induced oxidative stress and toxicity. *BioMed Research International, 2013*, 942916.

Mao, B. H., Tsai, J. C., Chen, C. W., Yan, S. J., & Wang, Y. J. (2016). Mechanisms of silver nanoparticle-induced toxicity and important role of autophagy. *Nanotoxicology, 10*(8), 1021–1040.

Meanwell, N. A. (2016). Improving drug design: An update on recent applications of efficiency metrics, strategies for replacing problematic elements, and compounds in nontraditional drug space. *Chemical Research in Toxicology, 29*(4), 564–616.

Meng, H., Leong, W., Leong, K. W., Chen, C., & Zhao, Y. (2018). Walking the line: The fate of nanomaterials at biological barriers. *Biomaterials, 174*, 41–53.

Mignani, S., Rodrigues, J., Tomas, H., Jalal, R., Singh, P. P., Majoral, J. P., & Vishwakarma, R. A. (2018). Present drug-likeness filters in medicinal chemistry during the hit and lead optimization process: How far can they be simplified? *Drug Discovery Today, 23*(3), 605–615.

Mirkin, C. A., Nel, A., & Thaxton, C. S. (2011). Applications: Nanobiosystems, medicine, and health. In *Nanotechnology research directions for societal needs in 2020* (pp. 305–374). Dordrecht: Springer.

Mohammadi, M. R., Nojoomi, A., Mozafari, M., Dubnika, A., Inayathullah, M., & Rajadas, J. (2017). Nanomaterials engineering for drug delivery: A hybridization approach. *Journal of Materials Chemistry B, 5*(22), 3995–4018.

Mohs, R. C., & Greig, N. H. (2017). Drug discovery and development: Role of basic biological research. *Alzheimer's and Dementia: Translational Research and Clinical Interventions, 3*(4), 651–657.

Navya, P. N., & Daima, H. K. (2016). Rational engineering of physicochemical properties of nanomaterials for biomedical applications with nanotoxicological perspectives. *Nano Convergence, 3*(1), 1.

Neagu, M., Piperigkou, Z., Karamanou, K., Engin, A. B., Docea, A. O., Constantin, C., et al. (2017). Protein bio-corona: critical issue in immune nanotoxicology. *Archives of Toxicology, 91*(3), 1031–1048.

Nehoff, H., Parayath, N. N., Domanovitch, L., Taurin, S., & Greish, K. (2014). Nanomedicine for drug targeting: Strategies beyond the enhanced permeability and retention effect. *International Journal of Nanomedicine, 9*, 2539.

Nekoueian, K., Amiri, M., Sillanpää, M., Marken, F., Boukherroub, R., & Szunerits, S. (2019). Carbon-based quantum particles: An electroanalytical and biomedical perspective. *Chemical Society Reviews, 48*, 4281.

Oberdörster, G., Maynard, A., Donaldson, K., Castranova, V., Fitzpatrick, J., Ausman, K., et al. (2005). Principles for characterizing the potential human health effects from exposure to nanomaterials: Elements of a screening strategy. *Particle and Fibre Toxicology, 2*(1), 8.

Olğaç, A., Türe, A., Olğaç, S., & Möller, S. (2019). Cloud-based high throughput virtual screening in novel drug discovery. In *High-performance modelling and simulation for big data applications* (pp. 250–278). Cham: Springer.

Papich, M. G., & Martinez, M. N. (2015). Applying biopharmaceutical classification system (BCS) criteria to predict oral absorption of drugs in dogs: Challenges and pitfalls. *The AAPS Journal, 17*(4), 948–964.

Park, J. K., Utsumi, T., Seo, Y. E., Deng, Y., Satoh, A., Saltzman, W. M., & Iwakiri, Y. (2016). Cellular distribution of injected PLGA-nanoparticles in the liver. *Nanomedicine: Nanotechnology, Biology and Medicine, 12*(5), 1365–1374.

Patra, J. K., Das, G., Fraceto, L. F., Campos, E. V. R., del Pilar Rodriguez-Torres, M., Acosta-Torres, L. S., et al. (2018). Nano based drug delivery systems: Recent developments and future prospects. *Journal of Nanobiotechnology, 16*(1), 71.

Pattnaik, S., Swain, K., & Lin, Z. (2016). Graphene and graphene-based nanocomposites: Biomedical applications and biosafety. *Journal of Materials Chemistry B, 4*(48), 7813–7831.

Pelaz, B., Alexiou, C., Alvarez-Puebla, R. A., Alves, F., Andrews, A. M., Ashraf, S., et al. (2017). Diverse applications of nanomedicine. *ACS Nano, 11*(3), 2313.

Pietroiusti, A., Magrini, A., & Campagnolo, L. (2016). New frontiers in nanotoxicology: Gut microbiota/microbiome-mediated effects of engineered nanomaterials. *Toxicology and Applied Pharmacology, 299*, 90–95.

Pietroiusti, A., Bergamaschi, E., Campagna, M., Campagnolo, L., De Palma, G., Iavicoli, S., et al. (2017). The unrecognized occupational relevance of the interaction between engineered nanomaterials and the gastro-intestinal tract: A consensus paper from a multidisciplinary working group. *Particle and Fibre Toxicology, 14*(1), 47.

Pizzorno, J. (2015). The kidney dysfunction epidemic, part 1: Causes. *Integrative Medicine: A Clinician's Journal, 14*(6), 8.

Qiao, Y., Ping, Y., Zhang, H., Zhou, B., Liu, F., Yu, Y., et al. (2019). Laser-activatable CuS nanodots to treat multidrug-resistant bacteria and release copper ion to accelerate healing of infected chronic nonhealing wounds. *ACS Applied Materials and Interfaces, 11*(4), 3809–3822.

Rauscher, H., Rasmussen, K., & Sokull-Klüttgen, B. (2017). Regulatory aspects of nanomaterials in the EU. *Chemie Ingenieur Technik, 89*(3), 224–231.

Recordati, C., De Maglie, M., Bianchessi, S., Argentiere, S., Cella, C., Mattiello, S., et al. (2015). Tissue distribution and acute toxicity of silver after single intravenous administration in mice: Nano-specific and size-dependent effects. *Particle and Fibre Toxicology, 13*(1), 12.

Rinninella, E., Raoul, P., Cintoni, M., Franceschi, F., Miggiano, G. A. D., Gasbarrini, A., & Mele, M. C. (2019). What is the healthy gut microbiota composition? A changing ecosystem across age, environment, diet, and diseases. *Microorganisms, 7*(1), 14.

Rizvi, S. A., & Saleh, A. M. (2018). Applications of nanoparticle systems in drug delivery technology. *Saudi Pharmaceutical Journal, 26*(1), 64–70.

Roach, K. A., Stefaniak, A. B., & Roberts, J. R. (2019). Metal nanomaterials: Immune effects and implications of physicochemical properties on sensitization, elicitation, and exacerbation of allergic disease. *Journal of Immunotoxicology, 16*(1), 87–124.

Sarmento, B., Andrade, F., Silva, S. B. D., Rodrigues, F., das Neves, J., & Ferreira, D. (2012). Cell-based in vitro models for predicting drug permeability. *Expert Opinion on Drug Metabolism and Toxicology, 8*(5), 607–621.

Scheinberg, D. A., Grimm, J., Heller, D. A., Stater, E. P., Bradbury, M., & McDevitt, M. R. (2017). Advances in the clinical translation of nanotechnology. *Current Opinion in Biotechnology, 46*, 66–73.

Schenone, M., Dančík, V., Wagner, B. K., & Clemons, P. A. (2013). Target identification and mechanism of action in chemical biology and drug discovery. *Nature Chemical Biology, 9*(4), 232.

Sharma, D., & Hussain, C. M. (2018). Smart nanomaterials in pharmaceutical analysis. *Arabian Journal of Chemistry, 13*(1), 3319–3343.

Sharma, A., Madhunapantula, S. V., & Robertson, G. P. (2012). Toxicological considerations when creating nanoparticle-based drugs and drug delivery systems. *Expert Opinion on Drug Metabolism and Toxicology, 8*(1), 47–69.

Sharma, A. K., Jaiswal, S. K., Chaudhary, N., & Sharma, V. K. (2017). A novel approach for the prediction of species-specific biotransformation of xenobiotic/drug molecules by the human gut microbiota. *Scientific Reports, 7*(1), 9751.

Shukla, S. K., Shukla, S. K., Govender, P. P., & Giri, N. G. (2016). Biodegradable polymeric nanostructures in therapeutic applications: Opportunities and challenges. *RSC Advances, 6*(97), 94325–94351.

Silva, A. M., Alvarado, H. L., Abrego, G., Martins-Gomes, C., Garduño-Ramirez, M. L., García, M. L., et al. (2019). In vitro cytotoxicity of oleanolic/ursolic acids-loaded in PLGA nanoparticles in different cell lines. *Pharmaceutics, 11*(8), 362.

Singh, R., & Lillard, J. W., Jr. (2009). Nanoparticle-based targeted drug delivery. *Experimental and Molecular Pathology, 86*(3), 215–223.

Singh, A. P., Biswas, A., Shukla, A., & Maiti, P. (2019). Targeted therapy in chronic diseases using nanomaterial-based drug delivery vehicles. *Signal Transduction and Targeted Therapy, 4*(1), 1–21.

Sonnenburg, J. L., & Fischbach, M. A. (2011). Community health care: Therapeutic opportunities in the human microbiome. *Science Translational Medicine, 3*(78), 78ps12–78ps12.

Stockdale, T. P., Challinor, V. L., Lehmann, R. P., De Voss, J. J., & Blanchfield, J. T. (2019). Caco-2 monolayer permeability and stability of Chamaelirium luteum (False Unicorn) open-chain steroidal saponins. *ACS Omega, 4*(4), 7658–7666.

Su, C., Liu, Y., Li, R., Wu, W., Fawcett, J. P., & Gu, J. (2019). Absorption, distribution, metabolism and excretion of the biomaterials used in Nanocarrier drug delivery systems. *Advanced Drug Delivery Reviews, 143*, 97–114.

Tavares, A. J., Poon, W., Zhang, Y. N., Dai, Q., Besla, R., Ding, D., et al. (2017). Effect of removing Kupffer cells on nanoparticle tumor delivery. *Proceedings of the National Academy of Sciences, 114*(51), E10871–E10880.

Teleanu, D. M., Chircov, C., Grumezescu, A. M., Volceanov, A., & Teleanu, R. I. (2018). Blood-brain delivery methods using nanotechnology. *Pharmaceutics, 10*(4), 269.

Tian, S., Wang, J., Li, Y., Li, D., Xu, L., & Hou, T. (2015). The application of in silico drug-likeness predictions in pharmaceutical research. *Advanced Drug Delivery Reviews, 86*, 2–10.

Tsaioun, K., Blaauboer, B. J., & Hartung, T. (2016). Evidence-based absorption, distribution, metabolism, excretion (ADME) and its interplay with alternative toxicity methods. *Alternatives to Animal Experimentation: ALTEX, 33*(4), 343–358.

Tsoi, K. M., MacParland, S. A., Ma, X. Z., Spetzler, V. N., Echeverri, J., Ouyang, B., et al. (2016). Mechanism of hard-nanomaterial clearance by the liver. *Nature Materials, 15*(11), 1212.

Tubert-Brohman, I., Sherman, W., Repasky, M., & Beuming, T. (2013). Improved docking of polypeptides with glide. *Journal of Chemical Information and Modeling, 53*(7), 1689–1699.

Uchegbu, I. F., Schätzlein, A. G., Cheng, W. P., & Lalatsa, A. (Eds.). (2013). *Fundamentals of pharmaceutical nanoscience*. Springer Science and Business Media, Berlin/Heidelberg, Germany.

ud Din, F., Aman, W., Ullah, I., Qureshi, O. S., Mustapha, O., Shafique, S., & Zeb, A. (2017). Effective use of nanocarriers as drug delivery systems for the treatment of selected tumors. *International Journal of Nanomedicine, 12*, 7291.

Ulbrich, K., Hola, K., Subr, V., Bakandritsos, A., Tucek, J., & Zboril, R. (2016). Targeted drug delivery with polymers and magnetic nanoparticles: Covalent and noncovalent approaches, release control, and clinical studies. *Chemical Reviews, 116*, 5338.

Upadhyay, R. K. (2014). Drug delivery systems, CNS protection, and the blood Brain barrier. *BioMed Research International, 2014*, 869269.

Van De Waterbeemd, H., Smith, D. A., Beaumont, K., & Walker, D. K. (2001). Property-based design: Optimization of drug absorption and pharmacokinetics. *Journal of Medicinal Chemistry, 44*(9), 1313–1333.

Vega-Villa, K. R., Takemoto, J. K., Yáñez, J. A., Remsberg, C. M., Forrest, M. L., & Davies, N. M. (2008). Clinical toxicities of nanocarrier systems. *Advanced Drug Delivery Reviews, 60*(8), 929–938.

Vitorino, C. V. (2018). Nanomedicine: Principles, properties and regulatory issues. *Frontiers in Chemistry, 6*, 360.

Volkov, Y., McIntyre, J., & Prina-Mello, A. (2017). Graphene toxicity as a double-edged sword of risks and exploitable opportunities: A critical analysis of the most recent trends and developments. *2D Materials, 4*(2), 022001.

Watson, C. Y., Molina, R. M., Louzada, A., Murdaugh, K. M., Donaghey, T. C., & Brain, J. D. (2015). Effects of zinc oxide nanoparticles on Kupffer cell phagosomal motility, bacterial clearance, and liver function. *International Journal of Nanomedicine, 10*, 4173.

Wilson, I. D., & Nicholson, J. K. (2017). Gut microbiome interactions with drug metabolism, efficacy, and toxicity. *Translational Research, 179*, 204–222.

Wong, O. A., Hansen, R. J., Ni, T. W., Heinecke, C. L., Compel, W. S., Gustafson, D. L., & Ackerson, C. J. (2013). Structure–activity relationships for biodistribution, pharmacokinetics, and excretion of atomically precise nanoclusters in a murine model. *Nanoscale, 5*(21), 10525–10533.

Zhang, M. Q., & Wilkinson, B. (2007). Drug discovery beyond the 'rule-of-five'. *Current Opinion in Biotechnology, 18*(6), 478–488.

Zhang, X. Q., Xu, X., Bertrand, N., Pridgen, E., Swami, A., & Farokhzad, O. C. (2012). Interactions of nanomaterials and biological systems: Implications to personalized nanomedicine. *Advanced Drug Delivery Reviews, 64*(13), 1363–1384.

Zhang, Y., Bai, Y., Jia, J., Gao, N., Li, Y., Zhang, R., et al. (2014). Perturbation of physiological systems by nanoparticles. *Chemical Society Reviews, 43*(10), 3762–3809.

Zhang, J., Li, Y., Chen, S. S., Zhang, L., Wang, J., Yang, Y., et al. (2015). Systems pharmacology dissection of the anti-inflammatory mechanism for the medicinal herb Folium eriobotryae. *International Journal of Molecular Sciences, 16*(2), 2913–2941.

Zhang, Y. N., Poon, W., Tavares, A. J., McGilvray, I. D., & Chan, W. C. (2016). Nanoparticle–liver interactions: Cellular uptake and hepatobiliary elimination. *Journal of Controlled Release, 240*, 332–348.

Zhang, N., Ming-Yuan Wei, M., & Ma, Q. (2019). Nanomedicines: A potential treatment for blood disorder diseases. *Frontiers in Bioengineering and Biotechnology, 7*, 369.

Zhou, Y., Peng, Z., Seven, E. S., & Leblanc, R. M. (2018). Crossing the blood-brain barrier with nanoparticles. *Journal of Controlled Release, 270*, 290–303.

Zhu, D., Long, Q., Xu, Y., & Xing, J. (2019). Evaluating nanoparticles in preclinical research using microfluidic systems. *Micromachines, 10*(6), 414.

Zolnik, B. S., Gonzalez-Fernandez, A., Sadrieh, N., & Dobrovolskaia, M. A. (2010). Minireview: Nanoparticles and the immune system. *Endocrinology, 151*(2), 458–465.

Chapter 6
Antimicrobial Nanotechnology in Preventing the Transmission of Infectious Disease

There is a difficulty with only one person changing. People call that person a great saint or a great mystic or a great leader, and they say, 'Well, he's different from me – I could never do it.' What's wrong with most people is that they have this block – they feel they could never make a difference, and therefore, they never face the possibility, because it is too disturbing, too frightening
Space is not empty. It is full, a plenum as opposed to a vacuum, and is the ground for the existence of everything, including ourselves. The universe is not separate from this cosmic sea of energy
Thus, in scientific research, a great deal of our thinking is in terms of theories. The word 'theory' derives from the Greek 'theoria', which has the same root as 'theatre', in a word meaning 'to view' or 'to make a spectacle'. Thus, it might be said that a theory is primarily a form of insight, i.e. a way of looking at the world, and not a form of knowledge of how the world is
— David Bohm (1917–1992)

Abstract Infections acquired in the hospital are one of the greatest threats to public health by increasing the morbidity and mortality of affected patients. In this way, it is necessary to implement robust and innovative control measures that prevent the formation of biofilms and the dissemination inside hospital areas. Thus, nanotechnology offers innovative solutions through the design and development of nanosurfaces capable of reducing the transmission, virulence, and infectivity of pathogenic microorganisms within medical care. In this order of ideas, the methods and protocols for evaluating nanomaterials functionalized with bioactive molecules applied to surfaces and medical devices acquire radical importance in order to determine activity and safety. For this reason the objective of this chapter is to make a comprehensive analysis of this interesting approach, in order to develop alternatives for control and prevention of infectious disease with low toxicity and an adequate

J. Bueno, *Preclinical Evaluation of Antimicrobial Nanodrugs*, Nanotechnology in the Life Sciences, https://doi.org/10.1007/978-3-030-43855-5_6

safety margin, which allows the implementation of hospital environments more biosafety and with greater protection for communities.

6.1 Introduction

The use of antimicrobial nanotechnology as an effective tool in the control of the spread of infectious disease is a promising application to prevent the emergence of antimicrobial resistance and reduce the indiscriminate use of antibiotics (Boukherroub et al. 2016; Hemeg 2017; Wang et al. 2017; Sani and Ehsani 2018). Thus, the use of nanomaterials allows not only the development of new medicines but also new surfaces and medical devices with anti-infective protection, which will reduce the microbial load in the hospital environment (Ramasamy and Lee 2016; Labreure et al. 2019; Lee and Jun 2019; Makowski et al. 2019). Likewise, nanocomposites can be used as biocidal agents and sterilizers capable of eliminating all contaminating microorganisms from hospital care centers (Álvarez-Paino et al. 2017; Rodríguez-Hernández 2017; Peddinti et al. 2019). Nanotechnology also presents a promising field of action in the design of equipment for the protection of health personnel at risk exposed to infectious agents, which may have an impact on the control of pathogenic microorganisms (Zhu et al. 2014; Coelho and García Díez 2015; Saccucci et al. 2018). On the other hand, nanocoatings become a major alternative which will prevent the growth of biofilms in hospital surfaces such as niche and focus of nosocomial infection (Fig. 6.1) (Ma et al. 2012; Khatoon et al. 2018; Ramos et al. 2018; Subhadra et al. 2018). Similarly nanocoatings can be applied on surgical sutures which can have a great impact on postsurgical infections impact on (Edmiston et al. 2013; De Simone et al. 2014; Reinbold et al. 2017; Ciraldo et al. 2019). For the reasons stated, the objective of this chapter is to analyze the role of preclinical evaluation in the design and implementation of antimicrobial

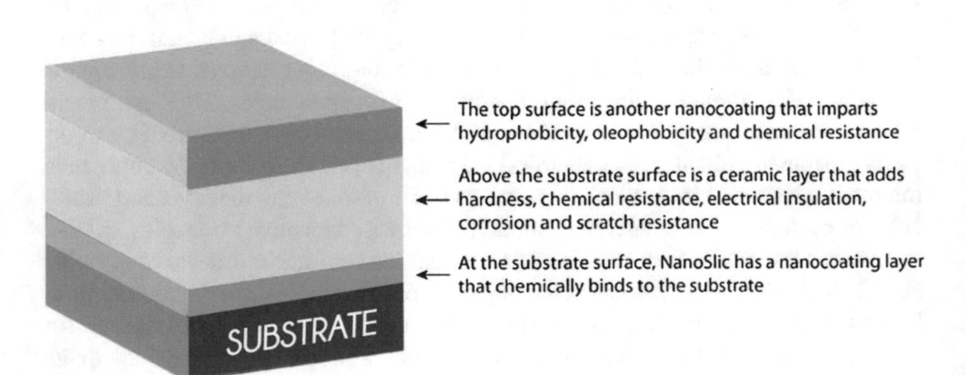

Fig. 6.1 Nanocoating structure

nanotechnology for the prevention and control of the spread of infectious disease from healthcare units to communities (Simpkin et al. 2017; Matteucci et al. 2018).

6.2 Nanomaterials in Antimicrobial Surfaces

Metal nanoparticles (Au, Cu, Zn, and Ag) have shown antimicrobial activity against gram-positive and gram-negative bacteria, so they are considered of importance to develop antimicrobial surfaces that prevent the adhesion of pathogens in the areas of care of the health (Beyth et al. 2015; Baptista et al. 2018; Hoseinnejad et al. 2018; Ruddaraju et al. 2019). Also nanotopography is a promising alternative capable of preventing the formation of microbial biofilms due to the transformation of the surface roughness of each nanocoating (Puckett et al. 2010; Besinis et al. 2017; Cheng et al. 2019). In the same way, it is possible the functionalization of the nanocoatings using antimicrobial molecules in combination with the nanomaterials within new nanostructures, which increase the anti-infective activity and the protection against biothreats inside the hospitals (Fig. 6.2) (Sampath Kumar and Madhumathi 2014; Francolini et al. 2017; Naskar and Kim 2019; Vazquez-Muñoz et al. 2019). Equally surgical masks, gloves, coats, surgical fields, and sutures can be functionalized with nanocomposites and biocides to design new devices that prevent the appearance of infection in the surgical site (Zhiqing et al. 2018; Genwa and Kumar 2019; Mariappan 2019; Patil et al. 2019). Theranostic nanostructures can also be used to develop new materials that allow detecting and treating microbial contamination from the infectious focus and thus avoiding the problem of infections in health personnel in specific cases of diseases with high transmissibility (Chen et al. 2014; Elsabahy et al. 2015; Ramasamy and Lee 2016; Martínez-Carmona et al. 2018). Thus, in this order of ideas, nanotechnology becomes one of the most promising approaches to prevent the transmission of infectious disease and infections associated with healthcare, allowing the development of hospital environments free of pathogens and increasing quality levels in clinical care (Okeke et al. 2011; Caliendo et al. 2013; Qasim et al. 2014; Chandler 2019).

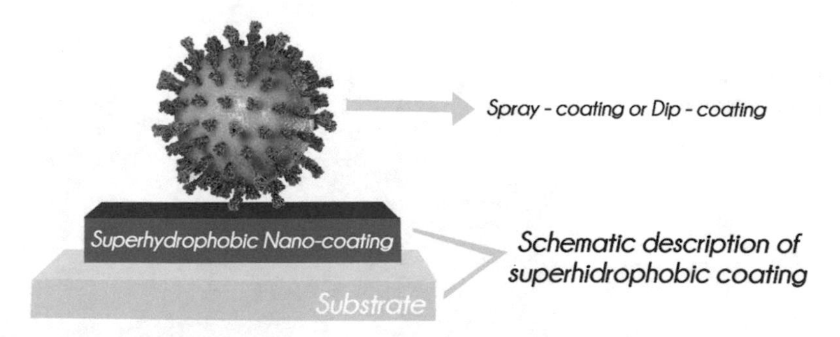

Fig. 6.2 Functionalized nanocoating

6.3 Antibiofilm Activity of Antimicrobial Nanotechnology

Biofilms have become the main focus of infection in hospital settings, as well as the niche of antimicrobial resistance in patients under clinical treatment in health-care centers (Adegoke et al. 2017; Frieri et al. 2017; Eze et al. 2018; D'Souza et al. 2019). In this order of ideas, antimicrobial nanotechnology allows the design of surfaces and devices capable of preventing the adhesion of sessile microorganisms that will subsequently form biofilms, avoiding the forms of microbial association that allow the exchange of genetic information and horizontal transmission of resistance (Borges et al. 2016; Natan and Banin 2017; Fuqua et al. 2019; Vallet-Regí et al. 2019). But perhaps the biggest challenge for nanomaterials is the elimination of the biofilm already established for which it is necessary to implement nano-tools capable of removing formations adhered to surfaces (Fig. 6.3) (Ostrikov et al. 2011; Liu et al. 2016; Qayyum and Khan 2016; Khelissa et al. 2017). On the other hand, biofilms as a formation for the survival and persistence of microbial cells have established resistance mechanisms such as low permeability and efflux pumps that decrease the concentration of antimicrobials, so it is of interest to design functionalized nano-films that inhibit resistance mechanisms, as well as virulence, reducing the emergence of microorganisms with infectious capacity (Lebeaux et al. 2014; Reygaert 2018; Lee et al. 2019; Reza et al. 2019). There are also nanomaterials that inhibit the production of acyl homoserine lactones which initiate the quorum sensing process that is the mechanism of cell grouping and communication for the formation of biofilms, making it a field of interest in the design of new antibiofilm surfaces (Gupta and Chhibber 2019; Husain et al. 2019; Jiang et al. 2019; López and Soto 2020).

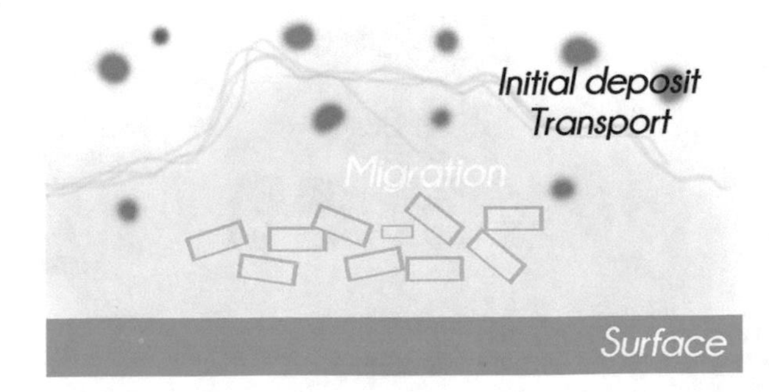

Fig. 6.3 Antibiofilm activity of nanomaterials

6.4 Antiquorum Sensing: Avoid Dissemination by Preventing Communication

Quorum sensing is one of the main phenomena of microbial physiology that mediates the communication, cell grouping, and formation of the biofilm matrix, so its inhibition is an effective therapeutic target in obtaining new antimicrobial nanosurfaces (Camele et al. 2019; Krzyżek 2019). Thus, in this order of ideas, the functionalization of an antibiotic film surface with molecules capable of preventing cell signaling processes, as well as the migration in microorganisms, is of great impact in the development of in-hospital environments and devices that reduce in-hospital infection (Mehrad et al. 2015; Yu et al. 2015; Riool et al. 2017; Zeng et al. 2018). In this way, carbon nanotubes, silver nanoparticles, and titanium dioxide have presented antiquorum sensing activity mediated by reactive oxygen species (ROS), which is why it is considered the basis for biomedical applications (Čáp et al. 2012; Fernandez-Bunster et al. 2012; Khanna et al. 2015; Mohanty et al. 2016). Thus, healthcare-associated infections are highly correlated with the presence of biofilms on environmental surfaces, both in clinical care facilities and in medical devices, so the evaluation of new materials that prevent the formation of microbial clusters by their surfactant action will be very useful to be used synergistically (Miquel et al. 2016; Adlhart et al. 2018; Magana et al. 2018; Rodrigues et al. 2020). Thus, the physicochemical properties of nanomaterials in interaction with biological systems allow the development of a new model of protection against bio-threats that requires new evaluation platforms that determine the antimicrobial activity and toxicity of surfaces, devices, and medical equipment.

6.5 Nanomaterials and Surfactant Activity

Thus nanomaterials have the ability to increase the surfactant activity of environmental surfaces in hospitals; this will result in further increased surface tension and physically inhibiting biofilm formation (Sadekuzzaman et al. 2015; Prasad et al. 2017; Koo et al. 2017; Satpute et al. 2019). Likewise, microorganisms will be more susceptible to antimicrobials due to the physical interruption of their interaction and communication in their sociomicrobiological activity in the formation of the biofilm; this will be very useful against the risk of polymicrobial biofilms where several species are in association for the persistence (Prasad et al. 2014; Roy et al. 2018; Ciofu and Tolker-Nielsen 2019; Gebreyohannes et al. 2019). In this order of ideas is possible to have inside the biofilm polymicrobial associations composed by bacteria and fungi which opens the way for a greater horizontal exchange of genetic information from prokaryotic to eukaryotic, which confers greater capacity for induction of antimicrobial resistance and more high rate of nosocomial infection, so a contamination control requires more suitable (Davies and Davies 2010; Giaouris et al. 2015; Juhas 2015; Rowan-Nash et al. 2019). Thus, it is in this aspect that the biophysical methods of eradication of biofilms that employ surface tension under the influence of nanomaterials can be an approximation of great impact on health-care environments and also decrease infections acquired in hospitals and the risk of health workers (Renner and Weibel 2011; Lorite et al. 2013; Song et al. 2015). Thus, the physicochemical properties of nanomaterials in interaction with biological systems allow the development of a new model of protection against biothreats that requires new evaluation platforms that determine the antimicrobial activity and toxicity of surfaces, devices, and medical equipment (Bohnsack et al. 2012; Cheng et al. 2013; Navya and Daima 2016; Burduşel et al. 2018).

6.6 Nano-Bio Interaction and Antimicrobial Effects

One of the main concerns of nanomaterials is to evaluate their antimicrobial activity on surfaces for the development of new control devices for the spread of infectious disease (Chang et al. 2019; Monteiro et al. 2019; Oves et al. 2019). Thus, methods and protocols have been established to determine the antimicrobial activity of nanoparticles arranged on a surface such as the coupon surfaces for evaluation of decontamination techniques proposed by the Environmental Protection Agency (EPA) in the EPA/600/R-12/591 protocol (Johns 2003; Calfee et al. 2012; Bloomfield et al. 2015; Velazquez et al. 2019). Likewise, the evaluation of nanotechnological compounds against biofilms has been implemented with flow tests such as Kadouri and modified Robbins device (Coenye et al. 2008; Bueno 2014; Otter et al. 2015; Günther et al. 2017). Flow biofilm assays are more predictive than static assays, because microbial formation is more resistant and various materials can be evaluated in which growth inhibition can be determined (Macia et al. 2014; Wilson et al. 2017; Bahamondez-Canas et al. 2019; Cattò and Cappitelli 2019). It is also possible to evaluate functionalized antimicrobial surfaces with nanostructures with the ASTM E2149-13a test methods (Standard Test Method for Determining the Antimicrobial Activity of Antimicrobial Agents Under Dynamic Contact Conditions) as a model for the evaluation of microorganism antiadhesion (Paladini et al. 2015; Campos et al. 2016; Caven et al. 2019; Lis et al. 2019). Finally, it is necessary to implement nanotoxicological models that evaluate the possible damages that antimicrobial surfaces may exert on health workers (Ahonen et al. 2017; Karahan et al. 2018; Montero et al. 2019).

6.7 Nanotoxicological Models in Antimicrobial Coatings

Thus, in this order of ideas, the evaluation models of the toxicity of antimicrobial surfaces require determining the risk of exposure to living organisms, so it is recommended to use whole animal models in the different tests such as zebrafish and *Galleria mellonella* (Megaw et al. 2015; Cassar et al. 2020; Cebrián et al. 2019;

Cutuli et al. 2019). These in vivo toxicity models have the advantage of being fast, economical, and robust in obtaining nanotoxicity results in target organs (Weichbrod et al. 2017). Likewise, toxicity can be determined in models that combine infectious diseases and antibiotic treatment, as well as the possibility of automating it using fluorescent markers, which makes them ideal and robust tests for nanomaterials (Novoa and Figueras 2012; Thomas et al. 2013; Cools et al. 2019). Likewise, the possibility of evaluating antibiotic film activity in these toxicity models is also a great advantage that zebrafish and *Galleria mellonella* can offer in order to determine if biocidal activity can cause systemic damage in living organisms (Ma et al. 2013; Tsai et al. 2016; Ignasiak and Maxwell 2017). Thus, in this order of ideas, the possibility of applying nanotechnology in care and protection in healthcare settings becomes more efficient as more information on safety is obtained to prevent toxicity (Seaton et al. 2009; Schulte et al. 2014).

6.8 Conclusions

The use of nanotechnology for the development of antimicrobial surfaces with which to reduce the incidence of hospital acquired infections is one of the great impacts of nanomaterials on public health (Weber et al. 2013; Grumezescu 2017; Arendsen et al. 2019; Muller et al. 2016). Thus, in this order of ideas, the application of nanoparticles in the different hospital surfaces becomes an effective mechanism to reduce microbial adhesion, biofilm formation, and the appearance of antimicrobial resistance within clinical areas (Konop et al. 2016). For this reason, the correct implementation of evaluation and safety protocols that determine the applicability of nanomaterials in healthcare facilities is required, in order to achieve a decrease in transmissibility and contamination to and from patients.

Acknowledgments The author thanks Sebastian Ritoré for his collaboration and invaluable support during the writing of this chapter, as well as the graphics contained in this book.

References

Adegoke, A., Faleye, A., Singh, G., & Stenström, T. (2017). Antibiotic resistant superbugs: Assessment of the interrelationship of occurrence in clinical settings and environmental niches. *Molecules, 22*(1), 29.

Adlhart, C., Verran, J., Azevedo, N. F., Olmez, H., Keinänen-Toivola, M. M., Gouveia, I., et al. (2018). Surface modifications for antimicrobial effects in the healthcare setting: A critical overview. *Journal of Hospital Infection, 99*(3), 239–249.

Ahonen, M., Kahru, A., Ivask, A., Kasemets, K., Kõljalg, S., Mantecca, P., et al. (2017). Proactive approach for safe use of antimicrobial coatings in healthcare settings: Opinion of the COST action network AMiCI. *International Journal of Environmental Research and Public Health, 14*(4), 366.

Álvarez-Paino, M., Muñoz-Bonilla, A., & Fernández-García, M. (2017). Antimicrobial polymers in the nano-world. *Nanomaterials, 7*(2), 48.

Arendsen, L. P., Thakar, R., & Sultan, A. H. (2019). The use of copper as an antimicrobial agent in health care, including obstetrics and gynecology. *Clinical Microbiology Reviews, 32*(4), e00125–e00118.

Bahamondez-Canas, T. F., Heersema, L. A., & Smyth, H. D. (2019). Current status of in vitro models and assays for susceptibility testing for wound biofilm infections. *Biomedicine, 7*(2), 34.

Baptista, P. V., McCusker, M. P., Carvalho, A., Ferreira, D. A., Mohan, N. M., Martins, M., & Fernandes, A. R. (2018). Nano-strategies to fight multidrug resistant bacteria—"A Battle of the Titans". *Frontiers in Microbiology, 9*, 1441.

Besinis, A., Hadi, S. D., Le, H. R., Tredwin, C., & Handy, R. D. (2017). Antibacterial activity and biofilm inhibition by surface modified titanium alloy medical implants following application of silver, titanium dioxide and hydroxyapatite nanocoatings. *Nanotoxicology, 11*(3), 327–338.

Beyth, N., Houri-Haddad, Y., Domb, A., Khan, W., & Hazan, R. (2015). Alternative antimicrobial approach: Nano-antimicrobial materials. *Evidence-Based Complementary and Alternative Medicine, 2015*. 15, 1, 42–59

Bloomfield, S., Exner, M., Flemming, H. C., Goroncy-Bermes, P., Hartemann, P., Heeg, P., et al. (2015). Lesser-known or hidden reservoirs of infection and implications for adequate prevention strategies: Where to look and what to look for. *GMS Hygiene and Infection Control, 10*, Doc04

Bohnsack, J. P., Assemi, S., Miller, J. D., & Furgeson, D. Y. (2012). The primacy of physicochemical characterization of nanomaterials for reliable toxicity assessment: A review of the zebrafish nanotoxicology model. In *Nanotoxicity* (pp. 261–316). Totowa: Humana Press.

Borges, A., Abreu, A., Dias, C., Saavedra, M., Borges, F., & Simões, M. (2016). New perspectives on the use of phytochemicals as an emergent strategy to control bacterial infections including biofilms. *Molecules, 21*(7), 877.

Boukherroub, R., Szunerits, S., & Drider, D. (Eds.). (2016). *Functionalized nanomaterials for the management of microbial infection: A strategy to address microbial drug resistance*. William Andrew, Norwich, NY.

Bueno, J. (2014). Anti-biofilm drug susceptibility testing methods: Looking for new strategies against resistance mechanism. *Journal of Microbial Biochemical Technology, 3*, 2.

Burduşel, A. C., Gherasim, O., Grumezescu, A., Mogoantă, L., Ficai, A., & Andronescu, E. (2018). Biomedical applications of silver nanoparticles: An up-to-date overview. *Nanomaterials, 8*(9), 681.

Calfee, M. W., Ryan, S. P., Wood, J. P., Mickelsen, L., Kempter, C., Miller, L., et al. (2012). Laboratory evaluation of large-scale decontamination approaches. *Journal of Applied Microbiology, 112*(5), 874–882.

Caliendo, A. M., Gilbert, D. N., Ginocchio, C. C., Hanson, K. E., May, L., Quinn, T. C., et al. (2013). Better tests, better care: Improved diagnostics for infectious diseases. *Clinical Infectious Diseases, 57*(suppl_3), S139–S170.

Camele, I., Elshafie, H. S., De Feo, V., & Caputo, L. (2019). Anti-quorum sensing and antimicrobial effect of mediterranean plant essential oils against phytopathogenic bacteria. *Frontiers in Microbiology, 10*, 2619.

Campos, M. D., Zucchi, P. C., Phung, A., Leonard, S. N., & Hirsch, E. B. (2016). The activity of antimicrobial surfaces varies by testing protocol utilized. *PLoS One, 11*(8), e0160728.

Čáp, M., Váchová, L., & Palková, Z. (2012). Reactive oxygen species in the signaling and adaptation of multicellular microbial communities. *Oxidative Medicine and Cellular Longevity, 2012*, 1.

Cassar, S., Adatto, I., Freeman, J. L., Gamse, J. T., Iturria, I., Lawrence, C., et al. (2020). Use of zebrafish in drug discovery toxicology. *Chemical Research in Toxicology, 33*(1), 95–118.

Cattò, C., & Cappitelli, F. (2019). Testing anti-biofilm polymeric surfaces: Where to start? *International Journal of Molecular Sciences, 20*(15), 3794.

Caven, B., Redl, B., & Bechtold, T. (2019). An investigation into the possible antibacterial properties of wool fibers. *Textile Research Journal, 89*(4), 510–516.

Cebrián, R., Rodríguez-Cabezas, M. E., Martín-Escolano, R., Rubiño, S., Garrido-Barros, M., Montalbán-López, M., et al. (2019). Preclinical studies of toxicity and safety of the AS-48 bacteriocin. *Journal of Advanced Research, 20*, 129–139.

Chandler, C. I. (2019). Current accounts of antimicrobial resistance: Stabilisation, individualisation and antibiotics as infrastructure. *Palgrave Communications, 5*(1), 53.

Chang, B. M., Pan, L., Lin, H. H., & Chang, H. C. (2019). Nanodiamond-supported silver nanoparticles as potent and safe antibacterial agents. *Scientific Reports, 9*(1), 1–11.

Chen, G., Qiu, H., Prasad, P. N., & Chen, X. (2014). Upconversion nanoparticles: Design, nanochemistry, and applications in theranostics. *Chemical Reviews, 114*(10), 5161–5214.

Cheng, L. C., Jiang, X., Wang, J., Chen, C., & Liu, R. S. (2013). Nano–bio effects: Interaction of nanomaterials with cells. *Nanoscale, 5*(9), 3547–3569.

Cheng, Y., Feng, G., & Moraru, C. I. (2019). Micro-and Nanotopography sensitive bacterial attachment mechanisms: A review. *Frontiers in Microbiology, 10*, 191.

Ciofu, O., & Tolker-Nielsen, T. (2019). Tolerance and resistance of Pseudomonas aeruginosa biofilms to antimicrobial agents-How P. aeruginosa can escape antibiotics. *Frontiers in Microbiology, 10*, 913.

Ciraldo, F. E., Schnepf, K., Goldmann, W. H., & Boccaccini, A. R. (2019). Development and characterization of bioactive glass containing composite coatings with ion releasing function for antibiotic-free antibacterial surgical sutures. *Materials, 12*(3), 423.

Coelho, A. C., & García Díez, J. (2015). Biological risks and laboratory-acquired infections: A reality that cannot be ignored in health biotechnology. *Frontiers in Bioengineering and Biotechnology, 3*, 56.

Coenye, T., De Prijck, K., De Wever, B., & Nelis, H. J. (2008). Use of the modified Robbins device to study the in vitro biofilm removal efficacy of NitrAdine™, a novel disinfecting formula for the maintenance of oral medical devices. *Journal of Applied Microbiology, 105*(3), 733–740.

Cools, F., Torfs, E., Aizawa Porto de Abreu, J., Vanhoutte, B., Maes, L., Caljon, G., et al. (2019). Optimization and characterization of a galleria mellonella larval infection model for virulence studies and the evaluation of therapeutics against Streptococcus pneumoniae. *Frontiers in Microbiology, 10*, 311.

Cutuli, M. A., Petronio Petronio, G., Vergalito, F., Magnifico, I., Pietrangelo, L., Venditti, N., & Di Marco, R. (2019). Galleria mellonella as a consolidated in vivo model hosts: New developments in antibacterial strategies and novel drug testing. *Virulence, 10*(1), 527–541.

D'Souza, A. W., Potter, R. F., Wallace, M., Shupe, A., Patel, S., Sun, X., et al. (2019). Spatiotemporal dynamics of multidrug resistant bacteria on intensive care unit surfaces. *Nature Communications, 10*(1), 1–19.

Davies, J., & Davies, D. (2010). Origins and evolution of antibiotic resistance. *Microbiology and Molecular Biology Reviews, 74*(3), 417–433.

De Simone, S., Gallo, A. L., Paladini, F., Sannino, A., & Pollini, M. (2014). Development of silver nano-coatings on silk sutures as a novel approach against surgical infections. *Journal of Materials Science: Materials in Medicine, 25*(9), 2205–2214.

Edmiston, C. E., Krepel, C. J., Marks, R. M., Rossi, P. J., Sanger, J., Goldblatt, M., et al. (2013). Microbiology of explanted suture segments from infected and noninfected surgical patients. *Journal of Clinical Microbiology, 51*(2), 417–421.

Elsabahy, M., Heo, G. S., Lim, S. M., Sun, G., & Wooley, K. L. (2015). Polymeric nanostructures for imaging and therapy. *Chemical Reviews, 115*(19), 10967–11011.

Eze, E. C., Chenia, H. Y., & El Zowalaty, M. E. (2018). Acinetobacter baumannii biofilms: Effects of physicochemical factors, virulence, antibiotic resistance determinants, gene regulation, and future antimicrobial treatments. *Infection and Drug Resistance, 11*, 2277.

Fernandez-Bunster, G., Gonzalez, C., Barros, J., & Martinez, M. (2012). Quorum sensing circuit and reactive oxygen species resistance in Deinococcus sp. *Current Microbiology, 65*(6), 719–725.

Francolini, I., Vuotto, C., Piozzi, A., & Donelli, G. (2017). Antifouling and antimicrobial biomaterials: An overview. *APMIS, 125*(4), 392–417.

Frieri, M., Kumar, K., & Boutin, A. (2017). Antibiotic resistance. *Journal of Infection and Public Health, 10*(4), 369–378.

Fuqua, C., Filloux, A., Ghigo, J. M., & Visick, K. L. (2019). Biofilms 2018: A diversity of microbes and mechanisms. *Journal of Bacteriology*, JB-00118.

Gebreyohannes, G., Nyerere, A., Bii, C., & Sbhatu, D. B. (2019). Challenges of intervention, treatment, and antibiotic resistance of biofilm-forming microorganisms. *Heliyon, 5*(8), e02192.

Genwa, M., & Kumar, P. (2019). Implications of nanotechnology in healthcare. *Nanoscience and Nanotechnology-Asia, 9*(1), 44–57.

Giaouris, E., Heir, E., Desvaux, M., Hebraud, M., Møretrø, T., Langsrud, S., et al. (2015). Intra-and inter-species interactions within biofilms of important foodborne bacterial pathogens. *Frontiers in Microbiology, 6*, 841.

Grumezescu, A. M. (Ed.). (2017). *Antimicrobial nanoarchitectonics: From synthesis to applications*. William Andrew, Norwich, NY.

Günther, F., Scherrer, M., Kaiser, S. J., DeRosa, A., & Mutters, N. T. (2017). Comparative testing of disinfectant efficacy on planktonic bacteria and bacterial biofilms using a new assay based on kinetic analysis of metabolic activity. *Journal of Applied Microbiology, 122*(3), 625–633.

Gupta, K., & Chhibber, S. (2019). Biofunctionalization of silver nanoparticles with lactonase leads to altered antimicrobial and cytotoxic properties. *Frontiers in Molecular Biosciences, 6*, 63.

Hemeg, H. A. (2017). Nanomaterials for alternative antibacterial therapy. *International Journal of Nanomedicine, 12*, 8211.

Hoseinnejad, M., Jafari, S. M., & Katouzian, I. (2018). Inorganic and metal nanoparticles and their antimicrobial activity in food packaging applications. *Critical Reviews in Microbiology, 44*(2), 161–181.

Husain, F. M., Ansari, A. A., Khan, A., Ahmad, N., Albadri, A., & Albalawi, T. H. (2019). Mitigation of acyl-homoserine lactone (AHL) based bacterial quorum sensing, virulence functions, and biofilm formation by yttrium oxide core/shell nanospheres: Novel approach to combat drug resistance. *Scientific Reports, 9*(1), 1–10.

Ignasiak, K., & Maxwell, A. (2017). Galleria mellonella (greater wax moth) larvae as a model for antibiotic susceptibility testing and acute toxicity trials. *BMC Research Notes, 10*(1), 428.

Jiang, Q., Chen, J., Yang, C., Yin, Y., & Yao, K. (2019). Quorum sensing: A prospective therapeutic target for bacterial diseases. *BioMed Research International, 2019*.

Johns, K. (2003). Hygienic coatings: The next generation. *Surface Coatings International Part B: Coatings Transactions, 86*(2), 101–110.

Juhas, M. (2015). Horizontal gene transfer in human pathogens. *Critical Reviews in Microbiology, 41*(1), 101–108.

Karahan, H. E., Wiraja, C., Xu, C., Wei, J., Wang, Y., Wang, L., et al. (2018). Graphene materials in antimicrobial nanomedicine: Current status and future perspectives. *Advanced Healthcare Materials, 7*(13), e1701406.

Khanna, P., Ong, C., Bay, B., & Baeg, G. (2015). Nanotoxicity: An interplay of oxidative stress, inflammation and cell death. *Nanomaterials, 5*(3), 1163–1180.

Khatoon, Z., McTiernan, C. D., Suuronen, E. J., Mah, T. F., & Alarcon, E. I. (2018). Bacterial biofilm formation on implantable devices and approaches to its treatment and prevention. *Heliyon, 4*(12), e01067.

Khelissa, S. O., Abdallah, M., Jama, C., Faille, C., & Chihib, N. E. (2017). Bacterial contamination and biofilm formation on abiotic surfaces and strategies to overcome their persistence. *Journal of Materials and Environmental Science, 8*, 3326–3346.

Konop, M., Damps, T., Misicka, A., & Rudnicka, L. (2016). Certain aspects of silver and silver nanoparticles in wound care: A minireview. *Journal of Nanomaterials, 2016*, 47.

Koo, H., Allan, R. N., Howlin, R. P., Stoodley, P., & Hall-Stoodley, L. (2017). Targeting microbial biofilms: Current and prospective therapeutic strategies. *Nature Reviews Microbiology, 15*(12), 740.

Krzyżek, P. (2019). Challenges and limitations of anti-quorum sensing therapies. *Frontiers in Microbiology, 10*, 2473.

Labreure, R., Sona, A. J., & Turos, E. (2019). Anti-methicillin resistant Staphylococcus aureus (MRSA) nanoantibiotics. *Frontiers in Pharmacology, 10*, 1121.

Lebeaux, D., Ghigo, J. M., & Beloin, C. (2014). Biofilm-related infections: Bridging the gap between clinical management and fundamental aspects of recalcitrance toward antibiotics. *Microbiology and Molecular Biology Reviews, 78*(3), 510–543.

Lee, S. H., & Jun, B. H. (2019). Silver nanoparticles: Synthesis and application for nanomedicine. *International Journal of Molecular Sciences, 20*(4), 865.

Lee, N. Y., Hsueh, P. R., & Ko, W. C. (2019). Nanoparticles in the treatment of infections caused by multidrug-resistant organisms. *Frontiers in Pharmacology, 10*, 1153.

Lis, M. J., Caruzi, B. B., Gil, G. A., Samulewski, R. B., Bail, A., Scacchetti, F. A. P., et al. (2019). In-situ direct synthesis of HKUST-1 in wool fabric for the improvement of antibacterial properties. *Polymers, 11*(4), 713.

Liu, S., Gunawan, C., Barraud, N., Rice, S. A., Harry, E. J., & Amal, R. (2016). Understanding, monitoring, and controlling biofilm growth in drinking water distribution systems. *Environmental Science and Technology, 50*(17), 8954–8976.

López, Y., & Soto, S. M. (2020). The usefulness of microalgae compounds for preventing biofilm infections. *Antibiotics, 9*(1), 9.

Lorite, G. S., Janissen, R., Clerici, J. H., Rodrigues, C. M., Tomaz, J. P., Mizaikoff, B., et al. (2013). Surface physicochemical properties at the micro and nano length scales: Role on bacterial adhesion and Xylella fastidiosa biofilm development. *PLoS One, 8*(9), e75247.

Ma, Y., Chen, M., Jones, J. E., Ritts, A. C., Yu, Q., & Sun, H. (2012). Inhibition of Staphylococcus epidermidis biofilm by trimethylsilane plasma coating. *Antimicrobial Agents and Chemotherapy, 56*(11), 5923–5937.

Ma, H., Williams, P. L., & Diamond, S. A. (2013). Ecotoxicity of manufactured ZnO nanoparticles–a review. *Environmental Pollution, 172*, 76–85.

Macia, M. D., Rojo-Molinero, E., & Oliver, A. (2014). Antimicrobial susceptibility testing in biofilm-growing bacteria. *Clinical Microbiology and Infection, 20*(10), 981–990.

Magana, M., Sereti, C., Ioannidis, A., Mitchell, C. A., Ball, A. R., Magiorkinis, E., et al. (2018). Options and limitations in clinical investigation of bacterial biofilms. *Clinical Microbiology Reviews, 31*(3), e00084–e00016.

Makowski, M., Silva, Í. C., Pais do Amaral, C., Gonçalves, S., & Santos, N. C. (2019). Advances in lipid and metal nanoparticles for antimicrobial peptide delivery. *Pharmaceutics, 11*(11), 588.

Mariappan, N. (2019). Recent trends in nanotechnology applications in surgical specialties and orthopedic surgery. *Biomedical and Pharmacology Journal, 12*(3), 1095–1127.

Martínez-Carmona, M., Gun'ko, Y., & Vallet-Regí, M. (2018). ZnO nanostructures for drug delivery and theranostic applications. *Nanomaterials, 8*(4), 268.

Matteucci, F., Giannantonio, R., Calabi, F., Agostiano, A., Gigli, G., and Rossi, M. (2018). Deployment and exploitation of nanotechnology nanomaterials and nanomedicine. In AIP conference proceedings (Vol. 1990, no. 1, p. 020001). AIP Publishing, College Park, Maryland.

Megaw, J., Thompson, T. P., Lafferty, R. A., & Gilmore, B. F. (2015). Galleria mellonella as a novel in vivo model for assessment of the toxicity of 1-alkyl-3-methylimidazolium chloride ionic liquids. *Chemosphere, 139*, 197–201.

Mehrad, B., Clark, N. M., Zhanel, G. G., & Lynch, J. P., III. (2015). Antimicrobial resistance in hospital-acquired gram-negative bacterial infections. *Chest, 147*(5), 1413–1421.

Miquel, S., Lagrafeuille, R., Souweine, B., & Forestier, C. (2016). Anti-biofilm activity as a health issue. *Frontiers in Microbiology, 7*, 592.

Mohanty, A., Tan, C. H., & Cao, B. (2016). Impacts of nanomaterials on bacterial quorum sensing: Differential effects on different signals. *Environmental Science: Nano, 3*(2), 351–356.

Monteiro, C., Costa, F., Pirttilä, A. M., Tejesvi, M. V., & Martins, M. C. L. (2019). Prevention of urinary catheter-associated infections by coating antimicrobial peptides from crowberry endophytes. *Scientific Reports, 9*(1), 1–14.

Montero, D. A., Arellano, C., Pardo, M., Vera, R., Gálvez, R., Cifuentes, M., et al. (2019). Antimicrobial properties of a novel copper-based composite coating with potential for use in healthcare facilities. *Antimicrobial Resistance and Infection Control, 8*(1), 3.

Muller, M. P., MacDougall, C., Lim, M., Armstrong, I., Bialachowski, A., Callery, S., et al. (2016). Antimicrobial surfaces to prevent healthcare-associated infections: A systematic review. *Journal of Hospital Infection, 92*(1), 7–13.

Naskar, A., & Kim, K. S. (2019). Nanomaterials as delivery vehicles and components of new strategies to combat bacterial infections: Advantages and limitations. *Microorganisms, 7*(9), 356.

Natan, M., & Banin, E. (2017). From nano to micro: Using nanotechnology to combat microorganisms and their multidrug resistance. *FEMS Microbiology Reviews, 41*(3), 302–322.

Navya, P. N., & Daima, H. K. (2016). Rational engineering of physicochemical properties of nanomaterials for biomedical applications with nanotoxicological perspectives. *Nano Convergence, 3*(1), 1.

Novoa, B., & Figueras, A. (2012). Zebrafish: Model for the study of inflammation and the innate immune response to infectious diseases. In *Current topics in innate immunity II* (pp. 253–275). New York: Springer.

Okeke, I. N., Peeling, R. W., Goossens, H., Auckenthaler, R., Olmsted, S. S., de Lavison, J. F., et al. (2011). Diagnostics as essential tools for containing antibacterial resistance. *Drug Resistance Updates, 14*(2), 95–106.

Ostrikov, K. K., Cvelbar, U., & Murphy, A. B. (2011). Plasma nanoscience: Setting directions, tackling grand challenges. *Journal of Physics D: Applied Physics, 44*(17), 174001.

Otter, J. A., Vickery, K., Walker, J. D., Pulcini, E. D., Stoodley, P., Goldenberg, S. D., et al. (2015). Surface-attached cells, biofilms and biocide susceptibility: Implications for hospital cleaning and disinfection. *Journal of Hospital Infection, 89*(1), 16–27.

Oves, M., Rauf, M. A., Qari, H. A., Muhammad, P., Khan, P. A., Ismail, I. M., et al. (2019). Antibacterial silver nanomaterials synthesis from Mesoflavibacter zeaxanthinifaciens and targeting biofilm formation. *Frontiers in Pharmacology, 10*, 801.

Paladini, F., Pollini, M., Sannino, A., & Ambrosio, L. (2015). Metal-based antibacterial substrates for biomedical applications. *Biomacromolecules, 16*(7), 1873–1885.

Patil, A., Mishra, V., Thakur, S., Riyaz, B., Kaur, A., Khursheed, R., et al. (2019). Nanotechnology derived nanotools in biomedical perspectives: An update. *Current Nanoscience, 15*(2), 137–146.

Peddinti, B. S., Scholle, F., Vargas, M. G., Smith, S. D., Ghiladi, R. A., & Spontak, R. J. (2019). Inherently self-sterilizing charged multiblock polymers that kill drug-resistant microbes in minutes. *Materials Horizons, 6*(10), 2056–2062.

Prasad, R., Shah, A. H., & Dhamgaye, S. (2014). Mechanisms of drug resistance in fungi and their significance in biofilms. In *Antibiofilm agents* (pp. 45–65). Berlin, Heidelberg: Springer.

Prasad, Y. S., Miryala, S., Lalitha, K., Ranjitha, K., Barbhaiwala, S., Sridharan, V., et al. (2017). Disassembly of bacterial biofilms by the self-assembled glycolipids derived from renewable resources. *ACS Applied Materials and Interfaces, 9*(46), 40047–40058.

Puckett, S. D., Taylor, E., Raimondo, T., & Webster, T. J. (2010). The relationship between the nanostructure of titanium surfaces and bacterial attachment. *Biomaterials, 31*(4), 706–713.

Qasim, M., Lim, D. J., Park, H., & Na, D. (2014). Nanotechnology for diagnosis and treatment of infectious diseases. *Journal of Nanoscience and Nanotechnology, 14*(10), 7374–7387.

Qayyum, S., & Khan, A. U. (2016). Nanoparticles vs. biofilms: A battle against another paradigm of antibiotic resistance. *MedChemComm, 7*(8), 1479–1498.

Ramasamy, M., & Lee, J. (2016). Recent nanotechnology approaches for prevention and treatment of biofilm-associated infections on medical devices. *BioMed Research International, 2016*, 1851242.

Ramos, M. A. D. S., Da Silva, P. B., Sposito, L., De Toledo, L. G., Bonifacio, B. V., Rodero, C. F., et al. (2018). Nanotechnology-based drug delivery systems for control of microbial biofilms: A review. *International Journal of Nanomedicine, 13*, 1179.

Reinbold, J., Uhde, A. K., Müller, I., Weindl, T., Geis-Gerstorfer, J., Schlensak, C., et al. (2017). Preventing surgical site infections using a natural, biodegradable, antibacterial coating on surgical sutures. *Molecules, 22*(9), 1570.

Renner, L. D., & Weibel, D. B. (2011). Physicochemical regulation of biofilm formation. *MRS Bulletin, 36*(5), 347–355.

Reygaert, W. C. (2018). An overview of the antimicrobial resistance mechanisms of bacteria. AIMS microbiology. *AIMS Microbiology, 4*(3), 482–501.

Reza, A., Sutton, J. M., & Rahman, K. M. (2019). Effectiveness of efflux pump inhibitors as biofilm disruptors and resistance breakers in gram-negative (ESKAPEE) bacteria. *Antibiotics, 8*(4), 229.

Riool, M., de Breij, A., Drijfhout, J. W., Nibbering, P. H., & Zaat, S. A. (2017). Antimicrobial peptides in biomedical device manufacturing. *Frontiers in Chemistry, 5*, 63.

Rodrigues, M. E., Gomes, F., & Rodrigues, C. F. (2020). Candida spp./Bacteria mixed biofilms. *Journal of Fungi, 6*(1), 5.

Rodríguez-Hernández, J. (2017). Polymers against microorganisms. In *Polymers against microorganisms* (pp. 1–11). Cham: Springer.

Rowan-Nash, A. D., Korry, B. J., Mylonakis, E., & Belenky, P. (2019). Cross-domain and viral interactions in the microbiome. *Microbiology and Molecular Biology Reviews, 83*(1), e00044–e00018.

Roy, R., Tiwari, M., Donelli, G., & Tiwari, V. (2018). Strategies for combating bacterial biofilms: A focus on anti-biofilm agents and their mechanisms of action. *Virulence, 9*(1), 522–554.

Ruddaraju, L. K., Pammi, S. V. N., Padavala, V. S., & Kolapalli, V. R. M. (2019). A review on anti-bacterials to combat resistance: From ancient era of plants and metals to present and future perspectives of green nano technological combinations. *Asian Journal of Pharmaceutical Sciences, 15*(1), 42–59.

Saccucci, M., Bruni, E., Uccelletti, D., Bregnocchi, A., Sarto, M. S., Bossù, M., et al. (2018). Surface disinfections: Present and future. *Journal of Nanomaterials, 2018*.

Sadekuzzaman, M., Yang, S., Mizan, M. F. R., & Ha, S. D. (2015). Current and recent advanced strategies for combating biofilms. *Comprehensive Reviews in Food Science and Food Safety, 14*(4), 491–509.

Sampath Kumar, T. S., & Madhumathi, K. (2014). Antibacterial potential of nanobioceramics used as drug carriers. In *Handbook of bioceramics and biocomposites* (pp. 1–42). Springer, Berlin/Heidelberg, Germany.

Sani, M. A., & Ehsani, A. (2018). Nanoparticles and their antimicrobial properties against pathogens including bacteria, fungi, parasites and viruses. *Microbial Pathogenesis, 123*, 505–526.

Satpute, S. K., Mone, N. S., Das, P., Banat, I. M., & Banpurkar, A. G. (2019). Inhibition of pathogenic bacterial biofilms on PDMS based implants by L. acidophilus derived biosurfactant. *BMC Microbiology, 19*(1), 39.

Schulte, P. A., Geraci, C. L., Murashov, V., Kuempel, E. D., Zumwalde, R. D., Castranova, V., et al. (2014). Occupational safety and health criteria for responsible development of nanotechnology. *Journal of Nanoparticle Research, 16*(1), 2153.

Seaton, A., Tran, L., Aitken, R., & Donaldson, K. (2009). Nanoparticles, human health hazard and regulation. *Journal of the Royal Society Interface, 7*(suppl_1), S119–S129.

Simpkin, V. L., Renwick, M. J., Kelly, R., & Mossialos, E. (2017). Incentivising innovation in antibiotic drug discovery and development: Progress, challenges and next steps. *The Journal of Antibiotics, 70*(12), 1087.

Song, F., Koo, H., & Ren, D. (2015). Effects of material properties on bacterial adhesion and biofilm formation. *Journal of Dental Research, 94*(8), 1027–1034.

Subhadra, B., Kim, D., Woo, K., Surendran, S., & Choi, C. (2018). Control of biofilm formation in healthcare: Recent advances exploiting quorum-sensing interference strategies and multidrug efflux pump inhibitors. *Materials, 11*(9), 1676.

Thomas, R. J., Hamblin, K. A., Armstrong, S. J., Müller, C. M., Bokori-Brown, M., Goldman, S., et al. (2013). Galleria mellonella as a model system to test the pharmacokinetics and efficacy of antibiotics against Burkholderia pseudomallei. *International Journal of Antimicrobial Agents, 41*(4), 330–336.

Tsai, C. J. Y., Loh, J. M. S., & Proft, T. (2016). Galleria mellonella infection models for the study of bacterial diseases and for antimicrobial drug testing. *Virulence, 7*(3), 214–229.

Vallet-Regí, M., González, B., & Izquierdo-Barba, I. (2019). Nanomaterials as promising alternative in the infection treatment. *International Journal of Molecular Sciences, 20*(15), 3806.

Vazquez-Muñoz, R., Meza-Villezcas, A., Fournier, P. G. J., Soria-Castro, E., Juarez-Moreno, K., Gallego-Hernández, A. L., et al. (2019). Enhancement of antibiotics antimicrobial activity due to the silver nanoparticles impact on the cell membrane. *PLoS One, 14*(11), e0224904.

Velazquez, S., Griffiths, W., Dietz, L., Horve, P., Nunez, S., Hu, J., et al. (2019). From one species to another: A review on the interaction between chemistry and microbiology in relation to cleaning in the built environment. *Indoor Air, 29*(6), 880.

Wang, L., Hu, C., & Shao, L. (2017). The antimicrobial activity of nanoparticles: Present situation and prospects for the future. *International Journal of Nanomedicine, 12*, 1227.

Weber, D. J., Anderson, D., & Rutala, W. A. (2013). The role of the surface environment in healthcare-associated infections. *Current Opinion in Infectious Diseases, 26*(4), 338–344.

Weichbrod, R. H., Thompson, G. A. H., & Norton, J. N. (2017). *Management of animal care and use programs in research, education, and testing*. CRC Press, Boca Raton, Florida.

Wilson, C., Lukowicz, R., Merchant, S., Valquier-Flynn, H., Caballero, J., Sandoval, J., et al. (2017). Quantitative and qualitative assessment methods for biofilm growth: A mini-review. *Research and Reviews Journal of Engineering and Technology, 6*(4).

Yu, Q., Wu, Z., & Chen, H. (2015). Dual-function antibacterial surfaces for biomedical applications. *Acta Biomaterialia, 16*, 1–13.

Zeng, Q., Zhu, Y., Yu, B., Sun, Y., Ding, X., Xu, C., et al. (2018). Antimicrobial and antifouling polymeric agents for surface functionalization of medical implants. *Biomacromolecules, 19*(7), 2805–2811.

Zhiqing, L., Yongyun, C., Wenxiang, C., Mengning, Y., Yuanqing, M., Zhenan, Z., et al. (2018). Surgical masks as source of bacterial contamination during operative procedures. *Journal of Orthopaedic Translation, 14*, 57–62.

Zhu, X., Radovic-Moreno, A. F., Wu, J., Langer, R., & Shi, J. (2014). Nanomedicine in the management of microbial infection–overview and perspectives. *Nano Today, 9*(4), 478–498.

Chapter 7
Nanotechnology in the Discovery of New Antimicrobial Drugs: Is a New Scientific Revolution Possible?

Of course the word chaos is used in rather a vague sense by a lot of writers, but in physics it means a particular phenomenon, namely that in a nonlinear system the outcome is often indefinitely, arbitrarily sensitive to tiny changes in the initial condition
But I don't actually adopt the point of view that our subjective impression of free will, which is a kind of indeterminacy behavior, comes from quantum mechanical indeterminacy
If we look at the way the universe behaves, quantum mechanics gives us fundamental, unavoidable indeterminacy, so that alternative histories of the universe can be assigned probability
— Murray Gell-mann (1929–2019)

Abstract Antimicrobial resistance and lack of research on new antibiotics are part of the perfect storm prior to pre-antibiotic apocalypse, where antimicrobial drugs commonly used in hospitals will not be useful against infectious disease, which will increase mortality and morbidity in all susceptible populations. Given the times of global crisis, solutions and alternatives of great impact with great depth and scientific knowledge are necessary. This leads us to the need to innovate in the midst of danger in order to develop comprehensive tools capable of diagnosing infectious disease early, preventing its spread and achieving an effective personalized treatment. It is in this area where nanotechnology becomes the most useful application to obtain renewed power and formula nanoantibiotics capable of addressing the great public health problems of this time, because it promises to change all the physical and chemical configuration of the current pharmacological science. So in this vein, the aim of this chapter is to analyze the integration of nanomaterials, pharmacology, and biosensors in a theranostics approach that will be useful for modern scope of the new smart antimicrobial drugs.

J. Bueno, *Preclinical Evaluation of Antimicrobial Nanodrugs*, Nanotechnology in the Life Sciences, https://doi.org/10.1007/978-3-030-43855-5_7

7.1 Introduction

The resistance of microorganisms to antimicrobial drugs has become a perfect storm, where few innovations are made as the mortality rate increases (Fair and Tor 2014; Khan and Khan 2016; Friedman et al. 2016; Aslam et al. 2018). Likewise, the indiscriminate and improper use of antibiotics has led to low effectiveness and success in anti-infective treatments, so it becomes a focus of dissemination of resistance and the emergence of new microorganisms with greater virulence and lower sensitivity to antimicrobials (Oliver et al. 2011; Beceiro et al. 2013; Manyi-Loh et al. 2018; Cheng et al. 2019). In the same way the resistance mechanisms of microorganisms such as biofilms have evolved due to the use of biocides and compounds such as triclosan, which results in a higher rate of dissemination of cross-resistance with hospital drugs, which aggravates more the situation (Dancer 2014; Foster 2017; Jutkina et al. 2018; Konai et al. 2018). In this order of ideas, a new era of antibiotics is necessary where not only microorganisms are inhibited but also their resistance and virulence mechanisms; they also manage to control transmissibility and microbial fitness (Martínez and Baquero 2002; Hobman and Crossman 2015; Munita and Arias 2016; Vila et al. 2016). Thus, nanotechnology offers the tools to recover the potency of anti-infective drugs, as well as for the design and development of new control and prevention measures for infectious disease (Hammond 2017; Baptista et al. 2018; Rex et al. 2019; Yang et al. 2019). In this way an integral union between pharmacology and physics is inevitable for the design and obtaining of modern antimicrobial nanopharmaceuticals, as well as diagnostic techniques with theranostics approach (Ahmad et al. 2017; Ventola 2017; Madamsetty et al. 2019; Rampado et al. 2019). Thus, the objective of this chapter is to analyze the impact of nanotechnology for the development of a new antibiotic era with which to face the perfect storm that constitutes antimicrobial resistance (Lawrence and Jeyakumar 2013; Konai et al. 2018; Hibbitts and O'Leary 2018; Sultan et al. 2018).

7.2 Nanotechnology vs Resistome: The Struggle for Survival and Persistence

One of the most important challenges in the control of antimicrobial resistance is in the inhibition of the mechanisms of microbial persistence known as resistome (Fig. 7.1.); this modulates the responses of microorganisms against antibiotic aggression and induces survival through resistance (Medina and Pieper 2016; Li and Webster 2018; Reygaert 2018; Kraemer et al. 2019). Thus, within the processes regulated by the resistome are the thickening of the microbial cell wall, the efflux of antibiotics, the increase in cell replication, cell signaling by quorum sensing, the conformation of biofilms, and the horizontal transfer of genetic material (Soto 2013; Singh et al. 2017a, b; Gebreyohannes et al. 2019; Saxena et al. 2019). In this order

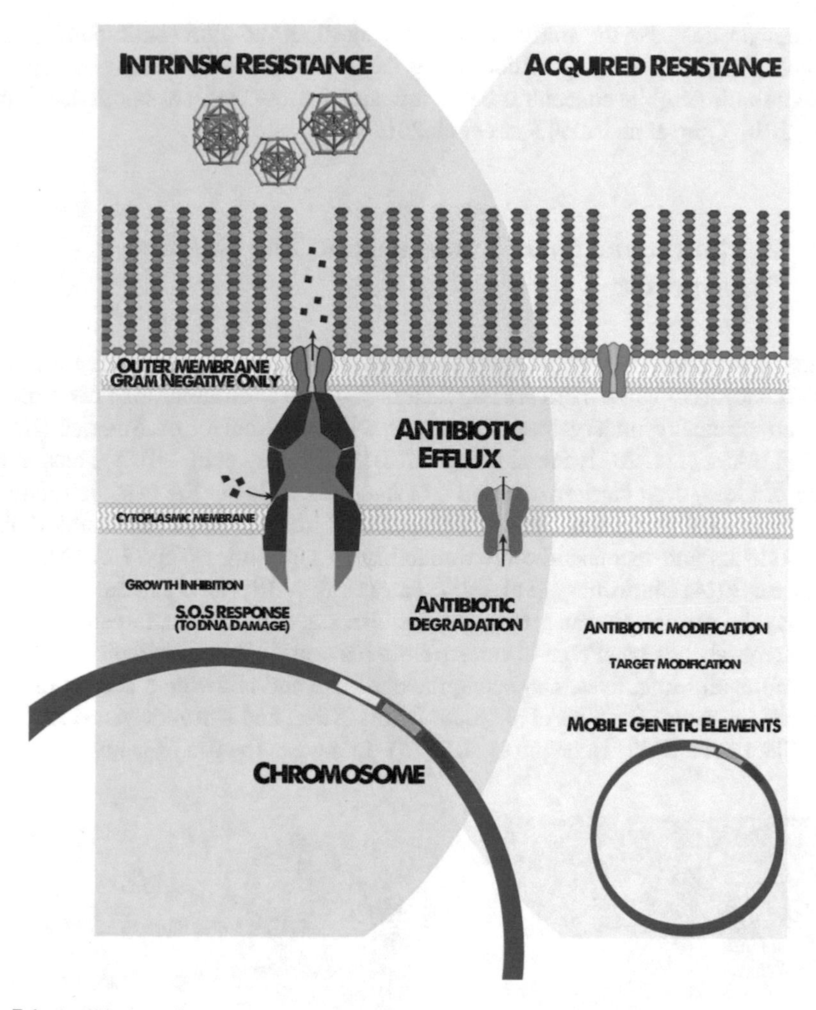

Fig. 7.1 Antibiotic resistome

of ideas, the inhibition of processes within the resistome by nanomaterials has been demonstrated, which is of importance in the development of antimicrobial adjuvants, capable of recovering the antibiotic power of drugs in microorganisms that have lost susceptibility (Wang et al. 2017; Flockton et al. 2019; Hu et al. 2019; Su et al. 2019). Likewise, nanostructures containing nanoparticles and functionalized with bioactive molecules have an interesting antibiotic activity, as well as inhibition of the mechanisms of persistence and microbial adaptation (Raghupathi et al. 2011; Rudramurthy et al. 2016; Sharma et al. 2017; Bodelón et al. 2018). Likewise, the blockage of efflux pumps as a mechanism to decrease the intracellular concentration of antibiotics is another of the therapeutic targets of nanomaterials that prevents the expulsion of antimicrobials from inside the microbial cell (Dreier and Ruggerone 2015; Shriram et al. 2018; Lowrence et al. 2019; Niño-Martínez et al. 2019). Finally, nanocomposites have the ability to permeate biofilms and allow antibiotics to enter the cells that make them up, which makes them effective drug transporters that can increase antimicrobial concentrations in infected tissues (Duncan et al. 2015; Ramos et al. 2018; Qing et al. 2019; Reza et al. 2019).

7.3 Antimicrobial Drug Nanocarriers: The Healing Transporter

Bringing antimicrobial medications to infected organs and maintaining adequate blood concentrations of them are two factors that lead to antimicrobial resistance as they are clinically undervalued (Infectious Diseases Society of America (IDSA) 2011; Leekha et al. 2011; Prestinaci et al. 2015; Hawkey et al. 2018). Thus, in this order of ideas, drug carriers composed of nanomaterials are the best alternative to cross biological barriers and allow access of antibiotics to areas of decreased pharmacokinetics and increase the bioavailability of antibiotics (Fig. 7.2) (Abed and Couvreur 2014; Chowdhury et al. 2017; Patra et al. 2018; Rizvi and Saleh 2018). In this way the use of liposomes, micelles, and nanoemulsions in combination with the antimicrobials has been able to improve the absorption and distribution in the tissues of the anti-infectives, increasing the cure rate and allowing a correct metabolism and excretion (Briones et al. 2008; Drulis-Kawa and Dorotkiewicz-Jach 2010; McMillan et al. 2011; Uchegbu et al. 2013). Likewise, the use of nanostructures in

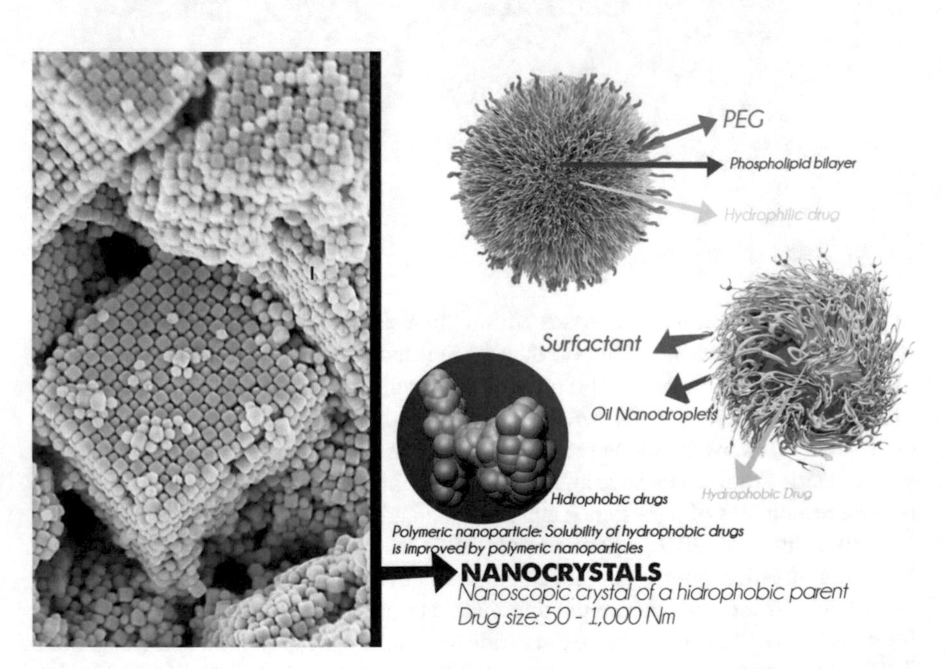

Fig. 7.2 Nanomaterials antimicrobial mechanism of action

a theranostics approach that allow diagnosis and timely treatment in the same compound is a promising alternative in drug carriers, which will reduce mortality from opportunistic infections in the future (Azmi and Shad 2017; Grumezescu 2017; Hoseinzadeh et al. 2017; Xie et al. 2017). On the other hand, the specificity of nanomaterials toward the target organ can be increased in association with specific ligands that recognize proteins from the affected tissues, as well as the microorganisms involved, which can allow the elimination of the pathogen without altering the microbiome (Docter et al. 2015; Biteen et al. 2016; Vidic et al. 2017; Dobrovolskaia 2019). Likewise, using the technique of specific ligands, therapeutic targets can be inhibited inside the biofilms, facilitating their eradication of both hospital environments and medical devices and tissues (Chung and Toh 2014; Ramasamy and Lee 2016; Koo et al. 2017; Jiao et al. 2019).

7.4 Intracellular Targets: Inhibition in the Specificity

One of the biggest promises of nanomaterials is their ability to inhibit therapeutic targets of microorganisms that are located intracellularly, which have greater latency and persistence, which makes them more difficult to treat and obtain an effective cure (Kamaruzzaman et al. 2017; Singh et al. 2017a, b; Ramirez-Acuña et al. 2019; Vallet-Regí et al. 2019). In this way the intracellular penetration inside the biofilms of the antimicrobials in order to kill dormant microorganisms specifically and effectively, as well as avoiding the acquisition of resistance by them, will be one of the greatest achievements of nanotechnology for the design and development of antibiotics (Lewis 2013; Mantravadi et al. 2019; Labreure et al. 2019; Van Giau et al. 2019). In this order of ideas, dormant microorganisms decrease their metabolic functions in order to persist the immune system and the antimicrobial activity of medications, which makes them very difficult to control especially in immunosuppressed patients (Fig. 7.3) (Cohen et al. 2013; Crucian et al. 2018; Pang et al. 2019). Thus, the most interesting intracellular therapeutic targets for nanotechnology are the respiratory chain, the expression of efflux pumps, and the horizontal transfer of genes, which not only boost persistence but also allow the transmissibility of resistance (Rai et al. 2013; Mulani et al. 2019; Mandal and Paul 2020; Ruddaraju et al. 2019; Mandal and Paul 2020). Also of concern is how to increase the inhibition of intracellular antimicrobial targets specifically without altering the physiological functioning of host tissues, in this way the design of nanostructures that are functionalized with specific antioxidant molecules and antibodies that recognize antigens from potentially pathogenic microbial is required (Natan and Banin 2017; Jahromi et al. 2018; Stewart et al. 2018; Manoharan et al. 2019). In this order of ideas, the use of aptamers such as short strands of DNA or RNA capable of binding with high affinity to target molecules is one of the approaches developed to be coupled to the nanocarriers and thus be able to increase the diagnostic and therapeutic potential of nanomaterials (Vorobyeva et al. 2016; Kalra et al. 2018; Sakai et al. 2018; Zhang et al. 2019). This leads us to the potentialities of the nanotheranostics

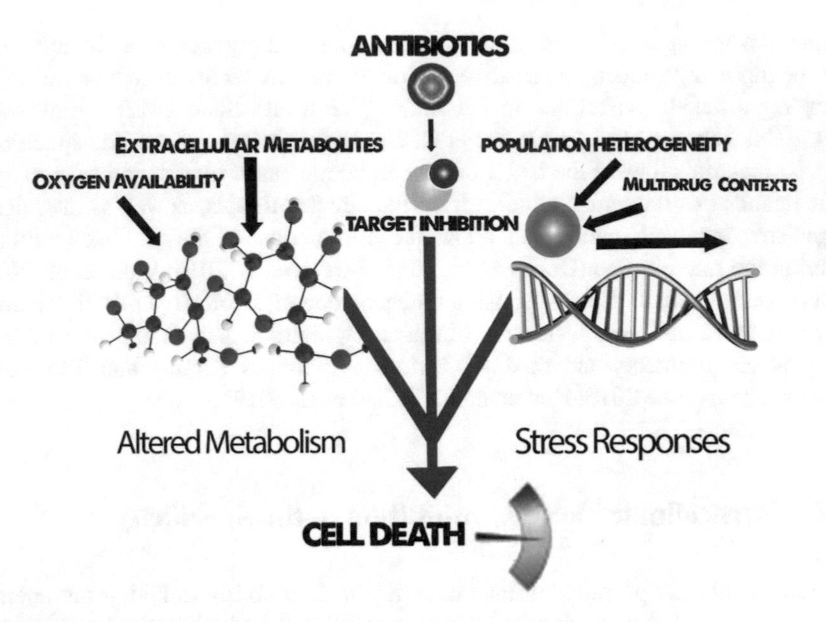

Fig. 7.3 Intracellular mechanisms of antimicrobial drug resistance

approach in which infectious disease can be diagnosed and treated using the same nanostructure, an approach that will allow infection control from any healthcare setting (Kim et al. 2013a, b; Jagtap et al. 2017; Colino et al. 2018; Bueno 2019).

7.5 Nanotheranostics in Infectious Diseases

The nanotheranostics approach consists in the development of multifunctional molecules composed of nanostructures associated with ligands as aptamers, as well as biomarkers and molecules with pharmacological activity, to perform biodiagnosis and treatment in the infectious focus (Patra et al. 2018; Taghdisi et al. 2018; Hosu et al. 2019; Lombardo et al. 2019). In this order of ideas, the coupling between microbial biomarkers, ligands, antibiotics, and fluorescent molecules in a complex nanostructure is a promising field for the development of the new nanotheranostics drugs (Kim et al. 2013a, b; Khlebtsov et al. 2013; Chen et al. 2014; Venditti 2019). These novel nanomedicaments will allow the early diagnosis of infectious diseases, as well as personalized treatment without affecting the microbiome and avoiding the use of broad-spectrum antibiotics, thereby reducing the risk of resistance (Langdon et al. 2016; Behrouzi et al. 2019; Gargiullo et al. 2019; Song et al. 2019). Thus, the development of nanotheranostics compounds requires the use of nanostructures and nanoscale objects to be functionalized such as nanosheets, nanopillars, nanorods, and nanodiamonds, among others; this allows greater interaction with the affected tissues and greater pharmacological activity

(Thanh et al. 2014; Sun et al. 2016; Perevedentseva et al. 2019; Song et al. 2019). Finally, this technology will open the doors for the development of modern nano-biosensors that will be useful in the discovery of new antimicrobial drugs with lower toxicity (Reder-Christ and Bendas 2011; Lan et al. 2017; Garzón et al. 2019; Pollap and Kochana 2019).

7.6 Nanobiosensors: Drug Discovery and Drug Monitoring

Thus, a promising application of nanomaterials is in the development of biosensors that can be used for the discovery of new medicines, as well as for the monitoring of the pharmacokinetics of drugs within the body (McKeating et al. 2016; Senapati et al. 2018; Vigneshvar and Senthilkumaran 2018; Zheng et al. 2019). In this way, nanobiosensors as antimicrobial drugs will allow obtaining information on the pathogen and infected tissue while giving relevant data on the presence of antimicrobial resistance and biofilms (Magana et al. 2018; Khatoon et al. 2018). Thus, medication and biodiagnosis are made in clinical care, which will be a great measure in public health for the detection of antimicrobial resistance, as well as for the elimination of contamination and dissemination foci (Prestinaci et al. 2015; Ayukekbong et al. 2017; Kraemer et al. 2019). On the other hand, the ability of biosensors to detect antibiotics in the environment will be of great help for the design and development of modern screening platforms that can determine the presence of new molecules in samples from biodiversity (Mehlhorn et al. 2018; Munteanu et al. 2018). Thus, the integration of nanomaterials into biosensors will lead us to a new era of intelligent medicines that accelerate diagnoses with personalized medical treatment (Yun et al. 2009; Prasad 2014; Martin et al. 2015).

7.7 Conclusions

The increase in antibiotic resistance and its dissemination will lead us to a pre-antibiotic apocalypse in which human beings do not have the capacity to control super microorganisms and succumb to them. Thus, in this order of ideas, the only solution to this threat to public health is the integral use of available technologies to develop viable solutions of great impact. So that is where nanotechnology can lead a scientific revolution that leads to a new antibiotic era that allows us to survive and persist. Finally, before the promising of nanotechnology, it is always necessary to remember that an integral model outside the pitfalls of disciplines is what is finally required for nanoantimicrobials, nanobiodiagnostic devices, and nano disinfection measurements to cross the valley of death in a translational science effort so that it can reach hospitals and healthcare units.

Acknowledgments The author thanks Sebastian Ritoré for his collaboration and invaluable support during the writing of this chapter, as well as the graphics contained in this book.

References

Abed, N., & Couvreur, P. (2014). Nanocarriers for antibiotics: A promising solution to treat intracellular bacterial infections. *International Journal of Antimicrobial Agents, 43*(6), 485–496.

Ahmad, J., Akhter, S., Rizwanullah, M., Ahmed Khan, M., Pigeon, L., Addo, R. T., et al. (2017). Nanotechnology based Theranostic approaches in Alzheimer's disease management: Current status and future perspective. *Current Alzheimer Research, 14*(11), 1164–1181.

Aslam, B., Wang, W., Arshad, M. I., Khurshid, M., Muzammil, S., Rasool, M. H., et al. (2018). Antibiotic resistance: A rundown of a global crisis. *Infection and Drug Resistance, 11*, 1645.

Ayukekbong, J. A., Ntemgwa, M., & Atabe, A. N. (2017). The threat of antimicrobial resistance in developing countries: Causes and control strategies. *Antimicrobial Resistance and Infection Control, 6*(1), 47.

Azmi, M. A., & Shad, K. F. (2017). Role of nanostructure molecules in enhancing the bioavailability of oral drugs. In *Nanostructures for novel therapy* (pp. 375–407). Elsevier. Amsterdam, Netherlands

Baptista, P. V., McCusker, M. P., Carvalho, A., Ferreira, D. A., Mohan, N. M., Martins, M., & Fernandes, A. R. (2018). Nano-strategies to fight multidrug resistant bacteria—"A Battle of the Titans". *Frontiers in Microbiology, 9*.

Beceiro, A., Tomás, M., & Bou, G. (2013). Antimicrobial resistance and virulence: A successful or deleterious association in the bacterial world? *Clinical Microbiology Reviews, 26*(2), 185–230.

Behrouzi, A., Nafari, A. H., & Siadat, S. D. (2019). The significance of microbiome in personalized medicine. *Clinical and Translational Medicine, 8*(1), 16.

Biteen, J. S., Blainey, P. C., Cardon, Z. G., Chun, M., Church, G. M., Dorrestein, P. C., et al. (2016). Tools for the microbiome: Nano and beyond. *ACS Nano, 10*(1), 6.

Bodelón, G., Montes-García, V., Pérez-Juste, J., & Pastoriza-Santos, I. (2018). Surface-enhanced Raman scattering spectroscopy for label-free analysis of P. aeruginosa quorum sensing. *Frontiers in Cellular and Infection Microbiology, 8*, 143.

Briones, E., Colino, C. I., & Lanao, J. M. (2008). Delivery systems to increase the selectivity of antibiotics in phagocytic cells. *Journal of Controlled Release, 125*(3), 210–227.

Bueno, J. (2019). Nanotheranostics approaches in antimicrobial drug resistance. In *Nanotheranostics* (pp. 41–61). Cham: Springer.

Chen, J., Wang, F., Liu, Q., & Du, J. (2014). Antibacterial polymeric nanostructures for biomedical applications. *Chemical Communications, 50*(93), 14482–14493.

Cheng, G., Ning, J., Ahmed, S., Huang, J., Ullah, R., An, B., et al. (2019). Selection and dissemination of antimicrobial resistance in Agri-food production. *Antimicrobial Resistance and Infection Control, 8*(1), 158.

Chowdhury, A., Kunjiappan, S., Panneerselvam, T., Somasundaram, B., & Bhattacharjee, C. (2017). Nanotechnology and nanocarrier-based approaches on treatment of degenerative diseases. *International Nano Letters, 7*(2), 91–122.

Chung, P. Y., & Toh, Y. S. (2014). Anti-biofilm agents: Recent breakthrough against multi-drug resistant Staphylococcus aureus. *Pathogens and Disease, 70*(3), 231–239.

Cohen, N. R., Lobritz, M. A., & Collins, J. J. (2013). Microbial persistence and the road to drug resistance. *Cell Host and Microbe, 13*(6), 632–642.

Colino, C., Millán, C., & Lanao, J. (2018). Nanoparticles for signaling in biodiagnosis and treatment of infectious diseases. *International Journal of Molecular Sciences, 19*(6), 1627.

Crucian, B. E., Choukèr, A., Simpson, R. J., Mehta, S., Marshall, G., Smith, S. M., et al. (2018). Immune system dysregulation during spaceflight: Potential countermeasures for deep space exploration missions. *Frontiers in Immunology, 9*, 1437.

Dancer, S. J. (2014). Controlling hospital-acquired infection: Focus on the role of the environment and new technologies for decontamination. *Clinical Microbiology Reviews, 27*(4), 665–690.

Dobrovolskaia, M. A. (2019). Nucleic acid nanoparticles at a crossroads of vaccines and immunotherapies. *Molecules, 24*(24), 4620.

Docter, D., Westmeier, D., Markiewicz, M., Stolte, S., Knauer, S. K., & Stauber, R. H. (2015). The nanoparticle biomolecule corona: Lessons learned–challenge accepted? *Chemical Society Reviews, 44*(17), 6094–6121.

Dreier, J., & Ruggerone, P. (2015). Interaction of antibacterial compounds with RND efflux pumps in Pseudomonas aeruginosa. *Frontiers in Microbiology, 6*, 660.

Drulis-Kawa, Z., & Dorotkiewicz-Jach, A. (2010). Liposomes as delivery systems for antibiotics. *International Journal of Pharmaceutics, 387*(1–2), 187–198.

Duncan, B., Li, X., Landis, R. F., Kim, S. T., Gupta, A., Wang, L. S., et al. (2015). Nanoparticle-stabilized capsules for the treatment of bacterial biofilms. *ACS Nano, 9*(8), 7775–7782.

Fair, R. J., & Tor, Y. (2014). Antibiotics and bacterial resistance in the 21st century. *Perspectives in Medicinal Chemistry, 6*, PMC-S14459.

Flockton, T. R., Schnorbus, L., Araujo, A., Adams, J., Hammel, M., & Perez, L. J. (2019). Inhibition of Pseudomonas aeruginosa biofilm formation with surface modified polymeric nanoparticles. *Pathogens, 8*(2), 55.

Foster, T. J. (2017). Antibiotic resistance in Staphylococcus aureus. Current status and future prospects. *FEMS Microbiology Reviews, 41*(3), 430–449.

Friedman, N. D., Temkin, E., & Carmeli, Y. (2016). The negative impact of antibiotic resistance. *Clinical Microbiology and Infection, 22*(5), 416–422.

Gargiullo, L., Del Chierico, F., D'Argenio, P., & Putignani, L. (2019). Gut microbiota modulation for multidrug-resistant organism decolonization: Present and future perspectives. *Frontiers in Microbiology, 10*, 1704.

Garzón, V., Pinacho, D. G., Bustos, R. H., Garzón, G., & Bustamante, S. (2019). Optical biosensors for therapeutic drug monitoring. *Biosensors, 9*(4), 132.

Gebreyohannes, G., Nyerere, A., Bii, C., & Sbhatu, D. B. (2019). Challenges of intervention, treatment, and antibiotic resistance of biofilm-forming microorganisms. *Heliyon, 5*(8), e02192.

Grumezescu, A. M. (Ed.). (2017). *Antimicrobial nanoarchitectonics: From synthesis to applications*. William Andrew, Norwich, NY.

Hammond, P. T. (2017). Nano tools pave the way to new solutions in infectious disease. *ACS Infectious Diseases, 3*(8), 554.

Hawkey, P. M., Warren, R. E., Livermore, D. M., McNulty, C. A., Enoch, D. A., Otter, J. A., & Wilson, A. P. R. (2018). Treatment of infections caused by multidrug-resistant gram-negative bacteria: Report of the British Society for Antimicrobial Chemotherapy/Healthcare Infection Society/British Infection Association Joint Working Party. *Journal of Antimicrobial Chemotherapy, 73*(suppl_3), iii2–iii78.

Hibbitts, A., & O'Leary, C. (2018). Emerging nanomedicine therapies to counter the rise of methicillin-resistant Staphylococcus aureus. *Materials, 11*(2), 321.

Hobman, J. L., & Crossman, L. C. (2015). Bacterial antimicrobial metal ion resistance. *Journal of Medical Microbiology, 64*, 471–497.

Hoseinzadeh, E., Makhdoumi, P., Taha, P., Hossini, H., Stelling, J., & Amjad Kamal, M. (2017). A review on nano-antimicrobials: Metal nanoparticles, methods and mechanisms. *Current Drug Metabolism, 18*(2), 120–128.

Hosu, O., Tertis, M., & Cristea, C. (2019). Implication of magnetic nanoparticles in cancer detection, screening and treatment. *Magnetochemistry, 5*(4), 55.

Hu, X., Yang, B., Zhang, W., Qin, C., Sheng, X., Oleszczuk, P., & Gao, Y. (2019). Plasmid binding to metal oxide nanoparticles inhibited lateral transfer of antibiotic resistance genes. *Environmental Science: Nano, 6*(5), 1310–1322.

Infectious Diseases Society of America (IDSA). (2011). Combating antimicrobial resistance: policy recommendations to save lives. *Clinical Infectious Diseases, 52*(suppl_5), S397–S428.

Jagtap, P., Sritharan, V., & Gupta, S. (2017). Nanotheranostic approaches for management of bloodstream bacterial infections. *Nanomedicine: Nanotechnology, Biology and Medicine, 13*(1), 329–341.

Jahromi, M. A. M., Zangabad, P. S., Basri, S. M. M., Zangabad, K. S., Ghamarypour, A., Aref, A. R., et al. (2018). Nanomedicine and advanced technologies for burns: Preventing infection and facilitating wound healing. *Advanced Drug Delivery Reviews, 123*, 33–64.

Jiao, Y., Tay, F. R., Niu, L. N., & Chen, J. H. (2019). Advancing antimicrobial strategies for managing oral biofilm infections. *International Journal of Oral Science, 11*(3), 1–11.

Jutkina, J., Marathe, N. P., Flach, C. F., & Larsson, D. G. J. (2018). Antibiotics and common antibacterial biocides stimulate horizontal transfer of resistance at low concentrations. *Science of the Total Environment, 616*, 172–178.

Kalra, P., Dhiman, A., Cho, W. C., Bruno, J. G., & Sharma, T. K. (2018). Simple methods and rational design for enhancing aptamer sensitivity and specificity. *Frontiers in Molecular Biosciences, 5*, 41.

Kamaruzzaman, N. F., Kendall, S., & Good, L. (2017). Targeting the hard to reach: Challenges and novel strategies in the treatment of intracellular bacterial infections. *British Journal of Pharmacology, 174*(14), 2225–2236.

Khan, S. N., & Khan, A. U. (2016). Breaking the spell: Combating multidrug resistant 'superbugs'. *Frontiers in Microbiology, 7*, 174.

Khatoon, Z., McTiernan, C. D., Suuronen, E. J., Mah, T. F., & Alarcon, E. I. (2018). Bacterial biofilm formation on implantable devices and approaches to its treatment and prevention. *Heliyon, 4*(12), e01067.

Khlebtsov, N., Bogatyrev, V., Dykman, L., Khlebtsov, B., Staroverov, S., Shirokov, A., et al. (2013). Analytical and theranostic applications of gold nanoparticles and multifunctional nanocomposites. *Theranostics, 3*(3), 167.

Kim, J. W., Galanzha, E. I., Zaharoff, D. A., Griffin, R. J., & Zharov, V. P. (2013a). Nanotheranostics of circulating tumor cells, infections and other pathological features in vivo. *Molecular Pharmaceutics, 10*(3), 813–830.

Kim, T. H., Lee, S., & Chen, X. (2013b). Nanotheranostics for personalized medicine. *Expert Review of Molecular Diagnostics, 13*(3), 257–269.

Konai, M. M., Bhattacharjee, B., Ghosh, S., & Haldar, J. (2018). Recent progress in polymer research to tackle infections and antimicrobial resistance. *Biomacromolecules, 19*(6), 1888–1917.

Koo, H., Allan, R. N., Howlin, R. P., Stoodley, P., & Hall-Stoodley, L. (2017). Targeting microbial biofilms: Current and prospective therapeutic strategies. *Nature Reviews Microbiology, 15*(12), 740.

Kraemer, S. A., Ramachandran, A., & Perron, G. G. (2019). Antibiotic pollution in the environment: From microbial ecology to public policy. *Microorganisms, 7*(6), 180.

Labreure, R., Sona, A. J., & Turos, E. (2019). Anti-methicillin resistant Staphylococcus aureus (MRSA) nanoantibiotics. *Frontiers in Pharmacology, 10*, 1121.

Lan, L., Yao, Y., Ping, J., & Ying, Y. (2017). Recent advances in nanomaterial-based biosensors for antibiotics detection. *Biosensors and Bioelectronics, 91*, 504–514.

Langdon, A., Crook, N., & Dantas, G. (2016). The effects of antibiotics on the microbiome throughout development and alternative approaches for therapeutic modulation. *Genome Medicine, 8*(1), 39.

Lawrence, R., & Jeyakumar, E. (2013, July). Antimicrobial resistance: A cause for global concern. In *BMC proceedings* (Vol. 7, no. 3, p. S1). BioMed Central, London, United Kingdom.

Leekha, S., Terrell, C. L., & Edson, R. S. (2011, February). General principles of antimicrobial therapy. In *Mayo clinic proceedings* (Vol. 86, no. 2, pp. 156-167). Elsevier. Amsterdam, Netherlands

Lewis, K. (2013). Platforms for antibiotic discovery. *Nature Reviews Drug Discovery, 12*(5), 371.

Li, B., & Webster, T. J. (2018). Bacteria antibiotic resistance: New challenges and opportunities for implant-associated orthopedic infections. *Journal of Orthopaedic Research®, 36*(1), 22–32.

Lombardo, D., Kiselev, M. A., & Caccamo, M. T. (2019). Smart nanoparticles for drug delivery application: Development of versatile nanocarrier platforms in biotechnology and nanomedicine. *Journal of Nanomaterials, 2019*, 1.

Lowrence, R. C., Subramaniapillai, S. G., Ulaganathan, V., & Nagarajan, S. (2019). Tackling drug resistance with efflux pump inhibitors: From bacteria to cancerous cells. *Critical Reviews in Microbiology, 45*(3), 334–353.

Madamsetty, V. S., Mukherjee, A., & Mukherjee, S. (2019). Recent trends of the bio-inspired nanoparticles in cancer theranostics. *Frontiers in Pharmacology, 10*, 1264.

Magana, M., Sereti, C., Ioannidis, A., Mitchell, C. A., Ball, A. R., Magiorkinis, E., et al. (2018). Options and limitations in clinical investigation of bacterial biofilms. *Clinical Microbiology Reviews, 31*(3), e00084–e00016.

Mandal, S. M., & Paul, D. (Eds.). (2020). *Bacterial adaptation to co-resistance*. Springer Nature. Berlin/Heidelberg, Germany

Manoharan, D., Li, W. P., & Yeh, C. S. (2019). Advances in controlled gas-releasing nanomaterials for therapeutic applications. *Nanoscale Horizons, 4*(3), 557–578.

Mantravadi, P. K., Kalesh, K. A., Dobson, R. C., Hudson, A. O., & Parthasarathy, A. (2019). The quest for novel antimicrobial compounds: Emerging trends in research, development, and technologies. *Antibiotics, 8*(1), 8.

Manyi-Loh, C., Mamphweli, S., Meyer, E., & Okoh, A. (2018). Antibiotic use in agriculture and its consequential resistance in environmental sources: Potential public health implications. *Molecules, 23*(4), 795.

Martin, J. H., Phillips, E., Thomas, D., & Somogyi, A. A. (2015). Adding the 'medicines' back into personalized medicine to improve cancer treatment outcomes. *British Journal of Clinical Pharmacology, 80*(5), 929.

Martínez, J. L., & Baquero, F. (2002). Interactions among strategies associated with bacterial infection: Pathogenicity, epidemicity, and antibiotic resistance. *Clinical Microbiology Reviews, 15*(4), 647–679.

McKeating, K. S., Aubé, A., & Masson, J. F. (2016). Biosensors and nanobiosensors for therapeutic drug and response monitoring. *Analyst, 141*(2), 429–449.

McMillan, J., Batrakova, E., & Gendelman, H. E. (2011). Cell delivery of therapeutic nanoparticles. In *Progress in molecular biology and translational science* (Vol. 104, pp. 563–601). Academic Press. Cambridge, Massachusetts

Medina, E., & Pieper, D. H. (2016). Tackling threats and future problems of multidrug-resistant bacteria. In *How to overcome the antibiotic crisis* (pp. 3–33). Cham: Springer.

Mehlhorn, A., Rahimi, P., & Joseph, Y. (2018). Aptamer-based biosensors for antibiotic detection: A review. *Biosensors, 8*(2), 54.

Mulani, M. S., Kamble, E. E., Kumkar, S. N., Tawre, M. S., & Pardesi, K. R. (2019). Emerging strategies to combat ESKAPE pathogens in the era of antimicrobial resistance: A review. *Frontiers in Microbiology, 10*, 539.

Munita, J. M., & Arias, C. A. (2016). Mechanisms of antibiotic resistance. *Microbiology Spectrum, 4*(2).

Munteanu, F. D., Titoiu, A., Marty, J. L., & Vasilescu, A. (2018). Detection of antibiotics and evaluation of antibacterial activity with screen-printed electrodes. *Sensors, 18*(3), 901.

Natan, M., & Banin, E. (2017). From nano to micro: Using nanotechnology to combat microorganisms and their multidrug resistance. *FEMS Microbiology Reviews, 41*(3), 302–322.

Niño-Martínez, N., Salas Orozco, M. F., Martínez-Castañón, G. A., Torres Méndez, F., & Ruiz, F. (2019). Molecular mechanisms of bacterial resistance to metal and metal oxide nanoparticles. *International Journal of Molecular Sciences, 20*(11), 2808.

Oliver, S. P., Murinda, S. E., & Jayarao, B. M. (2011). Impact of antibiotic use in adult dairy cows on antimicrobial resistance of veterinary and human pathogens: A comprehensive review. *Foodborne Pathogens and Disease, 8*(3), 337–355.

Pang, Z., Raudonis, R., Glick, B. R., Lin, T. J., & Cheng, Z. (2019). Antibiotic resistance in Pseudomonas aeruginosa: Mechanisms and alternative therapeutic strategies. *Biotechnology Advances, 37*(1), 177–192.

Patra, J. K., Das, G., Fraceto, L. F., Campos, E. V. R., del Pilar Rodriguez-Torres, M., Acosta-Torres, L. S., et al. (2018). Nano based drug delivery systems: Recent developments and future prospects. *Journal of Nanobiotechnology, 16*(1), 71.

Perevedentseva, E., Ali, N., Karmenyan, A., Skovorodkin, I., Prunskaite-Hyyryläinen, R., Vainio, S., et al. (2019). Optical studies of nanodiamond-tissue interaction: Skin penetration and localization. *Materials, 12*(22), 3762.

Pollap, A., & Kochana, J. (2019). Electrochemical immunosensors for antibiotic detection. *Biosensors, 9*(2), 61.

Prasad, S. (2014). Nanobiosensors: The future for diagnosis of disease. *Nanobiosensors in Disease Diagnosis, 3*, 1–10.

Prestinaci, F., Pezzotti, P., & Pantosti, A. (2015). Antimicrobial resistance: A global multifaceted phenomenon. *Pathogens and Global Health, 109*(7), 309–318.

Qing, G., Zhao, X., Gong, N., Chen, J., Li, X., Gan, Y., et al. (2019). Thermo-responsive triple-function nanotransporter for efficient chemo-photothermal therapy of multidrug-resistant bacterial infection. *Nature Communications, 10*(1), 1–12.

Raghupathi, K. R., Koodali, R. T., & Manna, A. C. (2011). Size-dependent bacterial growth inhibition and mechanism of antibacterial activity of zinc oxide nanoparticles. *Langmuir, 27*(7), 4020–4028.

Rai, S., Goel, S. K., Dwivedi, U. N., Sundar, S., & Goyal, N. (2013). Role of efflux pumps and intracellular thiols in natural antimony resistant isolates of Leishmania donovani. *PLoS One, 8*(9), e74862.

Ramasamy, M., & Lee, J. (2016). Recent nanotechnology approaches for prevention and treatment of biofilm-associated infections on medical devices. *BioMed Research International, 2016*, 1851242.

Ramirez-Acuña, J. M., Cardenas-Cadena, S. A., Marquez-Salas, P. A., Garza-Veloz, I., Perez-Favila, A., Cid-Baez, M. A., et al. (2019). Diabetic foot ulcers: Current advances in antimicrobial therapies and emerging treatments. *Antibiotics, 8*(4), 193.

Ramos, M. A. D. S., Da Silva, P. B., Sposito, L., De Toledo, L. G., Bonifacio, B. V., Rodero, C. F., et al. (2018). Nanotechnology-based drug delivery systems for control of microbial biofilms: A review. *International Journal of Nanomedicine, 13*, 1179.

Rampado, R., Crotti, S., Caliceti, P., Pucciarelli, S., & Agostini, M. (2019). Nanovectors design for theranostic applications in colorectal cancer. *Journal of Oncology, 2019*, 1.

Reder-Christ, K., & Bendas, G. (2011). Biosensor applications in the field of antibiotic research—A review of recent developments. *Sensors, 11*(10), 9450–9466.

Rex, J. H., Lynch, H. F., Cohen, I. G., Darrow, J. J., & Outterson, K. (2019). Designing development programs for non-traditional antibacterial agents. *Nature Communications, 10*(1), 1–10.

Reygaert, W. C. (2018). An overview of the antimicrobial resistance mechanisms of bacteria. *AIMS Microbiology, 4*(3), 482–501.

Reza, A., Sutton, J. M., & Rahman, K. M. (2019). Effectiveness of efflux pump inhibitors as biofilm disruptors and resistance breakers in gram-negative (ESKAPEE) Bacteria. *Antibiotics, 8*(4), 229.

Rizvi, S. A., & Saleh, A. M. (2018). Applications of nanoparticle systems in drug delivery technology. *Saudi Pharmaceutical Journal, 26*(1), 64–70.

Ruddaraju, L. K., Pammi, S. V. N., Padavala, V. S., & Kolapalli, V. R. M. (2019). A review on anti-bacterials to combat resistance: From ancient era of plants and metals to present and future perspectives of green nano technological combinations. *Asian Journal of Pharmaceutical Sciences, 15*(1), 42–59.

Rudramurthy, G., Swamy, M., Sinniah, U., & Ghasemzadeh, A. (2016). Nanoparticles: Alternatives against drug-resistant pathogenic microbes. *Molecules, 21*(7), 836.

Sakai, Y., Islam, M., Adamiak, M., Shiu, S. C. C., Tanner, J. A., & Heddle, J. G. (2018). DNA aptamers for the functionalisation of DNA origami nanostructures. *Genes, 9*(12), 571.

Saxena, P., Joshi, Y., Rawat, K., & Bisht, R. (2019). Biofilms: Architecture, resistance, quorum sensing and control mechanisms. *Indian Journal of Microbiology, 59*(1), 3–12.

Senapati, S., Mahanta, A. K., Kumar, S., & Maiti, P. (2018). Controlled drug delivery vehicles for cancer treatment and their performance. *Signal Transduction and Targeted Therapy, 3*(1), 7.

Sharma, C., Dhiman, R., Rokana, N., & Panwar, H. (2017). Nanotechnology: An untapped resource for food packaging. *Frontiers in Microbiology, 8*, 1735.

Shriram, V., Khare, T., Bhagwat, R., Shukla, R., & Kumar, V. (2018). Inhibiting bacterial drug efflux pumps via phyto-therapeutics to combat threatening antimicrobial resistance. *Frontiers in Microbiology, 9*.

Singh, L., Kruger, H. G., Maguire, G. E., Govender, T., & Parboosing, R. (2017a). The role of nanotechnology in the treatment of viral infections. *Therapeutic Advances in Infectious Disease, 4*(4), 105–131.

Singh, S., Singh, S. K., Chowdhury, I., & Singh, R. (2017b). Understanding the mechanism of bacterial biofilms resistance to antimicrobial agents. *The Open Microbiology Journal, 11*, 53.

Song, W., Anselmo, A. C., & Huang, L. (2019). Nanotechnology intervention of the microbiome for cancer therapy. *Nature Nanotechnology, 14*(12), 1093–1103.

Soto, S. M. (2013). Role of efflux pumps in the antibiotic resistance of bacteria embedded in a biofilm. *Virulence, 4*(3), 223–229.

Stewart, M. P., Langer, R., & Jensen, K. F. (2018). Intracellular delivery by membrane disruption: Mechanisms, strategies, and concepts. *Chemical Reviews, 118*(16), 7409–7531.

Su, Y., Wu, D., Xia, H., Zhang, C., Shi, J., Wilkinson, K. J., & Xie, B. (2019). Metallic nanoparticles induced antibiotic resistance genes attenuation of leachate culturable microbiota: The combined roles of growth inhibition, ion dissolution and oxidative stress. *Environment International, 128*, 407–416.

Sultan, I., Rahman, S., Jan, A. T., Siddiqui, M. T., Mondal, A. H., & Haq, Q. M. R. (2018). Antibiotics, resistome and resistance mechanisms: A bacterial perspective. *Frontiers in Microbiology, 9*.

Sun, Y., Yang, Q., & Wang, H. (2016). Synthesis and characterization of nanodiamond reinforced chitosan for bone tissue engineering. *Journal of Functional Biomaterials, 7*(3), 27.

Taghdisi, S. M., Danesh, N. M., Ramezani, M., Yazdian-Robati, R., & Abnous, K. (2018). A novel AS1411 aptamer-based three-way junction pocket DNA nanostructure loaded with doxorubicin for targeting cancer cells in vitro and in vivo. *Molecular Pharmaceutics, 15*(5), 1972–1978.

Thanh, N. T., Maclean, N., & Mahiddine, S. (2014). Mechanisms of nucleation and growth of nanoparticles in solution. *Chemical Reviews, 114*(15), 7610–7630.

Uchegbu, I. F., Schätzlein, A. G., Cheng, W. P., & Lalatsa, A. (Eds.). (2013). *Fundamentals of pharmaceutical nanoscience*. Springer Science and Business Media, Berlin/Heidelberg, Germany

Vallet-Regí, M., González, B., & Izquierdo-Barba, I. (2019). Nanomaterials as promising alternative in the infection treatment. *International Journal of Molecular Sciences, 20*(15), 3806.

Van Giau, V., An, S. S. A., & Hulme, J. (2019). Recent advances in the treatment of pathogenic infections using antibiotics and nano-drug delivery vehicles. *Drug Design, Development and Therapy, 13*, 327.

Venditti, I. (2019). Engineered gold-based nanomaterials: Morphologies and functionalities in biomedical applications. A mini review. *Bioengineering, 6*(2), 53.

Ventola, C. L. (2017). Progress in nanomedicine: Approved and investigational nanodrugs. *Pharmacy and Therapeutics, 42*(12), 742.

Vidic, J., Manzano, M., Chang, C. M., & Jaffrezic-Renault, N. (2017). Advanced biosensors for detection of pathogens related to livestock and poultry. *Veterinary Research, 48*(1), 11.

Vigneshvar, S., & Senthilkumaran, B. (2018). Current technological trends in biosensors, nanoparticle devices and biolabels: Hi-tech network sensing applications. *Medical Devices and Sensors, 1*(2), e10011.

Vila, J., Sáez-López, E., Johnson, J. R., Römling, U., Dobrindt, U., Cantón, R., et al. (2016). Escherichia coli: An old friend with new tidings. *FEMS Microbiology Reviews, 40*(4), 437–463.

Vorobyeva, M., Vorobjev, P., & Venyaminova, A. (2016). Multivalent aptamers: Versatile tools for diagnostic and therapeutic applications. *Molecules, 21*(12), 1613.

Wang, L., Hu, C., & Shao, L. (2017). The antimicrobial activity of nanoparticles: Present situation and prospects for the future. *International Journal of Nanomedicine, 12*, 1227.

Xie, S., Manuguri, S., Proietti, G., Romson, J., Fu, Y., Inge, A. K., et al. (2017). Design and synthesis of theranostic antibiotic nanodrugs that display enhanced antibacterial activity and luminescence. *Proceedings of the National Academy of Sciences, 114*(32), 8464–8469.

Yang, G., Chen, S., & Zhang, J. (2019). Bioinspired and biomimetic nanotherapies for the treatment of infectious diseases. *Frontiers in Pharmacology, 10*, 751.

Yun, Y. H., Eteshola, E., Bhattacharya, A., Dong, Z., Shim, J. S., Conforti, L., et al. (2009). Tiny medicine: Nanomaterial-based biosensors. *Sensors, 9*(11), 9275–9299.

Zhang, Y., Lai, B. S., & Juhas, M. (2019). Recent advances in aptamer discovery and applications. *Molecules, 24*(5), 941.

Zheng, F., Xiong, W., Sun, S., Zhang, P., & Zhu, J. J. (2019). Recent advances in drug release monitoring. *Nano, 8*(3), 391–413.

Chapter 8
Nanotechnology Beyond the Antibiosis

Science cannot be stopped. Man will gather knowledge no matter what the consequences – and we cannot predict what they will be. Science will go on — whether we are pessimistic, or are optimistic, as I am. I know that great, interesting, and valuable discoveries can be made and will be made... But I know also that still more interesting discoveries will be made that I have not the imagination to describe — and I am awaiting them, full of curiosity and enthusiasm

We may, I believe, anticipate that the chemist of the future who is interested in the structure of proteins, nucleic acids, polysaccharides, and other complex substances with high molecular weight will come to rely upon a new structural chemistry, involving precise geometrical relationships among the atoms in the molecules and the rigorous application of the new structural principles, and that great progress will be made, through this technique, in the attack, by chemical methods, on the problems of biology and medicine

What astonished me was the very low toxicity of a substance that has such very great physiological power. A little pinch, 5 mg, every day, is enough to keep a person from dying of pellagra, but it is so lacking in toxicity that ten thousand times as much can be taken without harm

— Linus Pauling (1901–1994)

Abstract The antibiotic era must pass to another stage if it wants to overcome antimicrobial resistance; this new phase or group of medications must have the ability to avoid the emergence of resistance while controlling infectious disease without altering the host's microbiome. In this order of ideas, the design and development of symbiotic medications such as those containing prebiotic, probiotic, and postbiotic metabolites are of great importance for the move toward a new antimicrobial model that takes symbiosis into account more than antibiosis, as a primary therapeutic target and that seeks to restore interactions between host and symbionts as part of the healing process. Thus, to implement this new approach, the available pharmacological tools are required in order to gain access to the largest number of host

J. Bueno, *Preclinical Evaluation of Antimicrobial Nanodrugs*, Nanotechnology in the Life Sciences, https://doi.org/10.1007/978-3-030-43855-5_8

microbiome species in order to develop multisystemic therapies that prevent dysbiosis, chronic inflammation, and other conditions such as cancer. Based on the foregoing, the objective of this chapter is to conduct a survey regarding modern nanoantimicrobials as part of a new symbiotic approach where the therapeutic target is the microbiome and its interactions, in order to modulate the restoration of both the immune system and the neuroendocrine.

8.1 Introduction

Thus, in the whole field of nanoantimicrobials, a new trend arises in the influence of nanomaterials on the human microbiome and the possibility of using nanotechnology for the design of prebiotic, probiotic, and symbiotic drugs that allow the maintenance and regulation of the microbiota in the tissues (Fig. 8.1) (Biteen et al. 2016; Karavolos and Holban 2016; Caneus 2017; Fadeel 2019). This new approach will allow the basis of the development of symbiotic drugs where it will be possible to couple nanomaterials with prebiotic molecules as well as polyphenols together with microorganisms that maintain the stability of the microbiome and thus achieve greater distribution in the mucous membranes (Siwek et al. 2018; Chong et al. 2019; Kumar Singh et al. 2019; Plichta et al. 2019). Based on the aforementioned evidence,

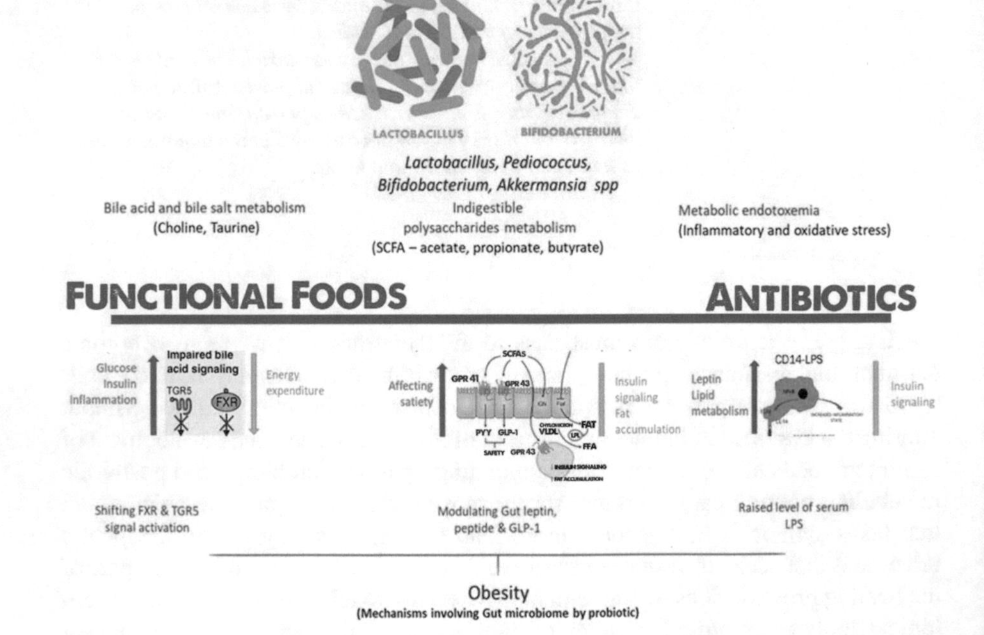

Fig. 8.1 Gut microbiome modulation by probiotics.jpeg

the accumulated data describes the maintenance of homeostasis and symbiosis in the interaction between the microbiome and the host as an important aspect to consider in the treatment of microbial infections, so nanotechnology should be applied as a tool effective to improve the administration of this type of medicines in the future (Chow et al. 2010; Hakansson et al. 2018; Sterlin et al. 2019; Zhang et al. 2019). Thus, the use of nanotechnology to balance the symbiotic holobiont (host + symbionts) may be the beginning of a modern antimicrobial therapy that keeps the immune system in competition and adaptive capacity (Singh et al. 2013; Dheilly 2014; Gjini and Brito 2016; Milani et al. 2017). In other words, the future of antimicrobial therapy will be more linked to the symbiosis of the holobiont than to the antibiosis of the pathogens that will determine an innovation in the acquisition of fitness and homeostasis by the human organism (Bordenstein and Theis 2015; Bang et al. 2018; Rosenberg and Zilber-Rosenberg 2018; Sitaraman 2018). For the above reasons, the objective of this chapter is to evaluate the applications of nanotechnology to symbiotic medicine as a starting point for a new era of infectious disease treatment (Chen et al. 2011; Jurj et al. 2017; Rauta et al. 2019).

8.2 Symbiotic Holobiont the Future of Therapy

Thus, the maintenance of homeostasis and symbiosis within the human holobiont is a therapeutic system that seeks to make interaction, communication, and exchange of information between the host and symbionts, the way for the biological system to continue adapting with entropy (Rosenberg et al. 2010; Foster et al. 2017; Macke et al. 2017). This causes the physiological functions of the host tissues in contact with the microbiome to remain in balance to keep both the immune and the neurological and endocrine systems active (Carabotti et al. 2015; Kho and Lal 2018; Ma et al. 2019; Zhang et al. 2019). In this order of ideas, symbiotic homeostasis allows the maintenance of tissue functionality and integrity, as well as the correct antigen recognition by the immune system, resulting in an adequate endocrine balance in the host (Belkaid and Hand 2014; Belkaid and Harrison 2017; Lin and Zhang 2017; De la Fuente 2018). It has also been shown that the correct homeostasis between the microbial species *Firmicutes*, *Bacteroides*, *Proteobacteria*, and *Actinobacteria* in the intestinal microbiome is the difference between a healthy patient and another in disease that presents microbiota alteration or dysbiosis (Fig. 8.2) (Thursby and Juge 2017; Huang and Shi 2019; Rinninella et al. 2019; Xu et al. 2019). Thus, dysbiosis becomes the biggest problem to be controlled in the symbiotic holobiont, so the influence of nanomaterials on this alteration must be evaluated and also determine which approaches would be adequate to enhance the restoration of microbial balance (Karavolos and Holban 2016; Poh et al. 2018; Qiu et al. 2018; Westmeier et al. 2018). Thus, in this way a strong correlation has been observed between dysbiosis, antimicrobial resistance, infectious disease, and the failure of antibiotic therapy, which makes the modulation of the microbiome a therapeutic target of interest in

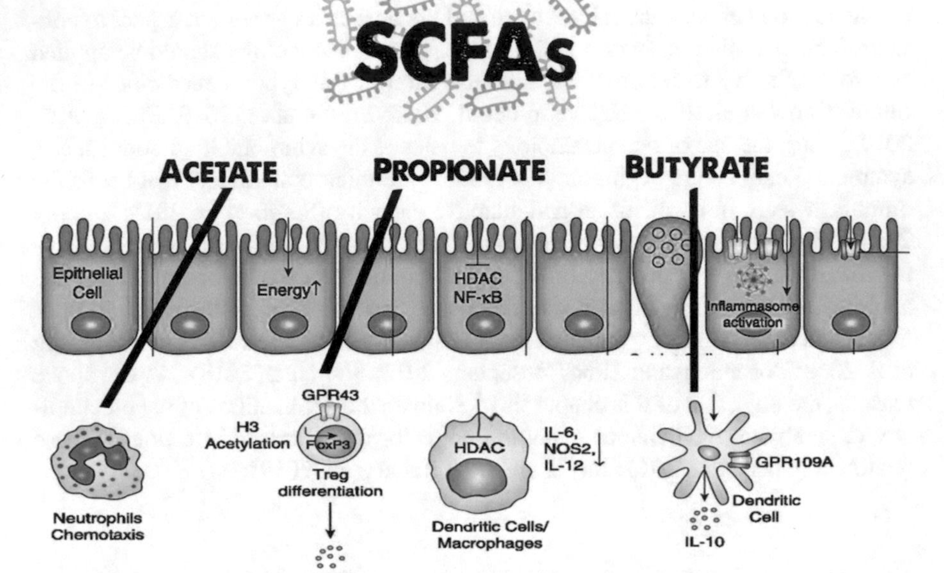

Fig. 8.2 Microbiome and immune system

the development of new medications (Becattini et al. 2016; Langdon et al. 2016; Galloway-Peña et al. 2017; Relman and Lipsitch 2018).

8.3 Dysbiosis and Nanomaterials: Potential Danger or a Great Opportunity?

Dysbiosis caused by the indiscriminate use of antibiotics and lifestyle is one of the greatest risks for the loss of symbiotic balance in the holobiont and induces chronic inflammation and antimicrobial resistance (Fig. 8.3) (Francino 2016; Kilian et al. 2016; Casals-Pascual et al. 2018; Neuman et al. 2018). Likewise, chronic inflammation induced by microbial imbalance is a factor that predisposes to cancer in affected populations, so it becomes a public health problem to highlight (Francescone et al. 2014; Morgillo et al. 2018; Pagliari et al. 2018; Baritaki et al. 2019; Hoare et al. 2019). For this reason it is important to determine the ability of nanoparticles to produce chronic inflammation through the alteration of the microbiome in dysbiosis, in order to establish reliable safety parameters (Hua et al. 2015; Rosenfeld 2017; Toribio-Mateas 2018; Siemer et al. 2018). In this order of ideas, it has been found in experimental animal models that silver nanoparticles alter the microbiome and neuronal behavior through the induction of dysbiosis (Javurek et al. 2017; Dahiya and Puniya 2018; Jiang et al. 2018; Qiu et al. 2018). But the protective effect of

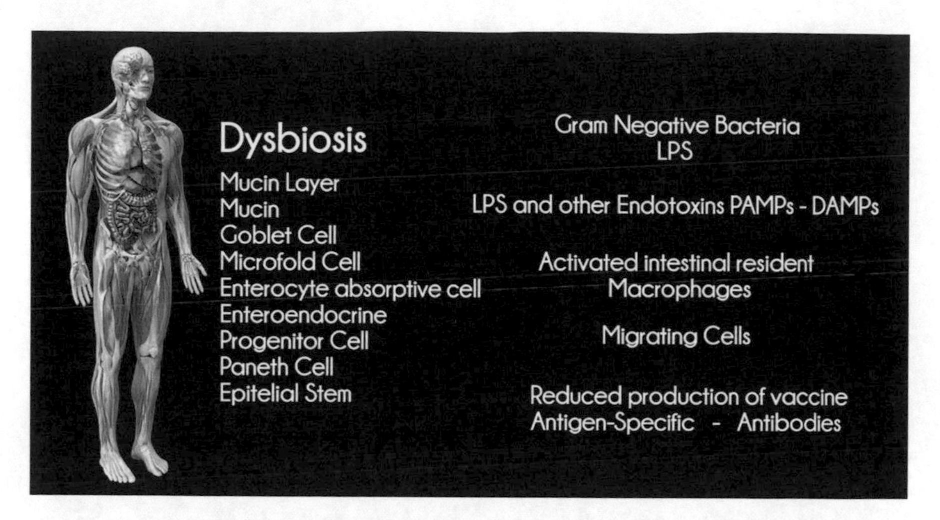

Fig. 8.3 Intestinal microbiome dysbiosis

polyphenol quercetin on tissue damage that metal nanoparticles can exert has also been reported, which opens the door for the development of nanomaterials associated with antioxidant molecules to reduce their toxicity (Khushnud and Mousa 2013; Mittal et al. 2014; Milinčić et al. 2019). On the other hand, nanocarriers such as liposomes have shown promise for encapsulation of probiotics and their metabolites in order to obtain new symbiotic medications useful in medicine (Anal and Singh 2007; Govender et al. 2014; Nguyen et al. 2017; Kerry et al. 2018). Thus, the prevention and control of dysbiosis will be part of a new field of research in the development of new drugs capable of reducing the incidence and prevalence of infectious diseases and cancer (McFarland 2014; Golemis et al. 2018; Saus et al. 2019; Tuteja and Ferguson 2019).

8.4 Nanocarriers and Symbiotic Drugs: Transporting Homeostasis

Thus nanocarriers such as liposomes, micelles, and polymers can transport prebiotic, probiotic, and postbiotic metabolites to different tissues in contact with the microbiome to maintain the balance and functioning of both immune and endocrine systems (Hemarajata and Versalovic 2013; Albillos et al. 2019; Cerdó et al. 2019; Jampilek et al. 2019). In this way, being able to systematically distribute the postbiotic metabolites to the entire human microbiome will allow restoring the altered symbiosis and maintaining the adequate exchange of information in the holobiont (Postler and Ghosh 2017; Selber-Hnatiw et al. 2017; Thomas et al. 2017; Cani 2018). Likewise, these postbiotic metabolites such as lactic acid and bacteriocins are of great interest to prevent the appearance of antimicrobial resistance and chronic infectious conditions such as those produced by *Clostridium difficile* (Ouwehand et al. 2016; Mokoena 2017; Van Giau et al. 2019; Vieco-Saiz et al. 2019). Thus, nanocarriers that transport symbiotic medications will achieve control of dysbiosis in order to reduce the indiscriminate use of antibiotics both in agriculture and in hospitals, which will result in a decrease in antimicrobial resistance and its fatal consequences (Dumitrescu et al. 2018; Chibbar and Dieleman 2019; Parker et al. 2020). On the other hand, drug carriers are of fundamental importance to be able to bring prebiotic, probiotic, and postbiotic metabolites into the biofilms in order to reduce their impact on infectious disease and public health (Miquel et al. 2016; Monteagudo-Mera et al. 2019; Terpou et al. 2019; Wegh et al. 2019).

8.5 Biofilms in Symbiosis

In this order of ideas, the biofilm constitutes a microbial formation that seeks persistence in the face of environmental aggression, be it ultraviolet light or antibiotics; thus, microorganisms to survive and evolve use defense and resistance mechanisms, and then infectious disease becomes an alteration of the human holobiont ecosystem (Dang and Lovell 2016; Skillings 2016; Yin et al. 2019; Simon et al. 2019). Thus, postbiotic metabolites and probiotic microorganisms have the ability to interact with the bacteria immersed in the biofilms by means of cellular signaling mechanisms and thus allow their symbiosis with the host (Salas-Jara et al. 2016; Salazar et al. 2016; Appanna 2018; Kimelman and Shemesh 2019). Similarly, probiotic microorganisms such as *Lactobacillus* can form biofilms spontaneously and establish communication with other species to establish polymicrobial associations in the microbiome (Lebeer et al. 2008; Jones and Versalovic 2009; Peters et al. 2012; Lamont et al. 2018). Thus, the restoration of homeostasis goes through the restoration of symbiosis within the biofilms that make up the microbiota, so that the exchange of genetic information in the hologenome and molecular interaction remain in equilibrium (Flandroy et al. 2018; Godoy-Vitorino 2019). Finally, the ability of nanomaterials coupled to medicines to enter the interior of biofilms will be the way to design drugs that go beyond antibiosis and enter a new stage of the management of infectious disease based on symbiosis (Lebeaux et al. 2014; Li et al. 2015; Nguyen and Hiorth 2015; Stanton et al. 2017).

8.6 Conclusions

The maintenance of homeostasis between the symbionts and the host inside the holobiont is a promising field of radical importance for the development of a modern antimicrobial therapy that prevents the development of antibiotic resistance (Haque and Haque 2017). In this order of ideas, nanotechnology is consolidated as the ideal encapsulation tool for the transport of prebiotic, probiotic, and postbiotic medications to the entire host microbiota for the restoration of communication and interaction, influencing both the immune system and the endocrine system. It is also necessary to evaluate the possible alteration of the human microbiome by nanomaterials which may cause dysbiosis, chronic inflammation, and cancer (Francescone et al. 2014; Ding et al. 2018). Thus, the integration of nanotechnology with symbiotic medicine will bring about the development of novel biotherapeutics that will reduce morbidity and mortality due to infectious diseases caused by multidrug-resistant microorganisms and cancer (Afolayan et al. 2018; Lee et al. 2019).

Acknowledgments The author thanks Sebastian Ritoré for his collaboration and invaluable support during the writing of this chapter, as well as the graphics contained in this book.

References

Afolayan, A. O., Adetoye, A., & Ayeni, F. A. (2018). Beneficial microbes: Roles in the era of antimicrobial resistance. In *Antimicrobial resistance-A global threat*. IntechOpen. Rijeka, Croatia

Albillos, A., Gottardi, A., & Rescigno, M. (2019). The gut-liver axis in liver disease: Pathophysiological basis for therapy. *Journal of Hepatology, 72*(3), 558–577

Anal, A. K., & Singh, H. (2007). Recent advances in microencapsulation of probiotics for industrial applications and targeted delivery. *Trends in Food Science and Technology, 18*(5), 240–251.

Appanna, V. D. (2018). *Human microbes-the power within: Health, healing and beyond*. Springer. Berlin/Heidelberg, Germany

Bang, C., Dagan, T., Deines, P., Dubilier, N., Duschl, W. J., Fraune, S., et al. (2018). Metaorganisms in extreme environments: Do microbes play a role in organismal adaptation? *Zoology, 127*, 1–19.

Baritaki, S., de Bree, E., Chatzaki, E., & Pothoulakis, C. (2019). Chronic stress, inflammation, and colon cancer: A CRH system-driven molecular crosstalk. *Journal of Clinical Medicine, 8*(10), 1669.

Becattini, S., Taur, Y., & Pamer, E. G. (2016). Antibiotic-induced changes in the intestinal microbiota and disease. *Trends in Molecular Medicine, 22*(6), 458–478.

Belkaid, Y., & Hand, T. W. (2014). Role of the microbiota in immunity and inflammation. *Cell, 157*(1), 121–141.

Belkaid, Y., & Harrison, O. J. (2017). Homeostatic immunity and the microbiota. *Immunity, 46*(4), 562–576.

Biteen, J. S., Blainey, P. C., Cardon, Z. G., Chun, M., Church, G. M., Dorrestein, P. C., et al. (2016). Tools for the microbiome: Nano and beyond. *ACS Nano, 10*(1), 6.

Bordenstein, S. R., & Theis, K. R. (2015). Host biology in light of the microbiome: Ten principles of holobionts and hologenomes. *PLoS Biology, 13*(8), e1002226.

Caneus, D. (2017). Nanotechnology and its partnership with synbiotics. *Journal of Nanomedicine Research, 6*(1), 00142.

Cani, P. D. (2018). Human gut microbiome: Hopes, threats and promises. *Gut, 67*(9), 1716–1725.

Carabotti, M., Scirocco, A., Maselli, M. A., & Severi, C. (2015). The gut-brain axis: Interactions between enteric microbiota, central and enteric nervous systems. *Annals of Gastroenterology: Quarterly Publication of the Hellenic Society of Gastroenterology, 28*(2), 203.

Casals-Pascual, C., Vergara, A., & Vila, J. (2018). Intestinal microbiota and antibiotic resistance: Perspectives and solutions. *Human Microbiome Journal, 9*, 11–15.

Cerdó, T., García-Santos, J. A., Bermúdez, M. G., & Campoy, C. (2019). The role of probiotics and prebiotics in the prevention and treatment of obesity. *Nutrients, 11*(3), 635.

Chen, T., Vargeese, C., Vagle, K., Wang, W., & Zhang, Y. (2011). U.S. Patent No. 7,893,302. Washington, DC: U.S. Patent and Trademark Office.

Chibbar, R., & Dieleman, L. A. (2019). The gut microbiota in celiac disease and probiotics. *Nutrients, 11*(10), 2375.

Chong, P. P., Chin, V. K., Looi, C. Y., Wong, W. F., Madhavan, P., & Yong, V. C. (2019). The microbiome and irritable bowel syndrome–a review on the pathophysiology, current research and future therapy. *Frontiers in Microbiology, 10*, 1136.

Chow, J., Lee, S. M., Shen, Y., Khosravi, A., & Mazmanian, S. K. (2010). Host–bacterial symbiosis in health and disease. In *Advances in immunology* (Vol. 107, pp. 243–274). Academic Press.

Dahiya, D. K., & Puniya, A. K. (2018). Impact of nanosilver on gut microbiota: A vulnerable link. *Future Microbiology, 13*(4), 483–492.

Dang, H., & Lovell, C. R. (2016). Microbial surface colonization and biofilm development in marine environments. *Microbiology and Molecular Biology Reviews, 80*(1), 91–138.

De la Fuente, M. (2018). Oxidation and inflammation in the immune and nervous systems, a link between aging and anxiety. In *Handbook of immunosenescence: Basic understanding and clinical implications* (pp. 1–31). Springer, Berlin/Heidelberg, Germany

Dheilly, N. M. (2014). Holobiont–holobiont interactions: Redefining host–parasite interactions. *PLoS Pathogens, 10*(7), e1004093.

Ding, C., Tang, W., Fan, X., & Wu, G. (2018). Intestinal microbiota: A novel perspective in colorectal cancer biotherapeutics. *Oncotargets and Therapy, 11*, 4797.

Dumitrescu, L., Popescu-Olaru, I., Cozma, L., Tulbă, D., Hinescu, M. E., Ceafalan, L. C., et al. (2018). Oxidative stress and the microbiota-gut-brain axis. *Oxidative Medicine and Cellular Longevity, 2018*, 1.

Fadeel, B. (2019). The right stuff: On the future of nanotoxicology. *Frontiers in Toxicology, 1*, 1.

Flandroy, L., Poutahidis, T., Berg, G., Clarke, G., Dao, M. C., Decaestecker, E., et al. (2018). The impact of human activities and lifestyles on the interlinked microbiota and health of humans and of ecosystems. *Science of the Total Environment, 627*, 1018–1038.

Foster, K. R., Schluter, J., Coyte, K. Z., & Rakoff-Nahoum, S. (2017). The evolution of the host microbiome as an ecosystem on a leash. *Nature, 548*(7665), 43.

Francescone, R., Hou, V., & Grivennikov, S. I. (2014). Microbiome, inflammation and cancer. *Cancer Journal (Sudbury, Mass.), 20*(3), 181.

Francino, M. P. (2016). Antibiotics and the human gut microbiome: Dysbioses and accumulation of resistances. *Frontiers in Microbiology, 6*, 1543.

Galloway-Peña, J. R., Jenq, R. R., & Shelburne, S. A. (2017). Can consideration of the microbiome improve antimicrobial utilization and treatment outcomes in the oncology patient? *Clinical Cancer Research, 23*(13), 3263–3268.

Gjini, E., & Brito, P. H. (2016). Integrating antimicrobial therapy with host immunity to fight drug-resistant infections: Classical vs. adaptive treatment. *PLoS Computational Biology, 12*(4), e1004857.

Godoy-Vitorino, F. (2019). Human microbial ecology and the rising new medicine. *Annals of Translational Medicine, 7*(14), 342.

Golemis, E. A., Scheet, P., Beck, T. N., Scolnick, E. M., Hunter, D. J., Hawk, E., & Hopkins, N. (2018). Molecular mechanisms of the preventable causes of cancer in the United States. *Genes and Development, 32*(13–14), 868–902.

Govender, M., Choonara, Y. E., Kumar, P., du Toit, L. C., van Vuuren, S., & Pillay, V. (2014). A review of the advancements in probiotic delivery: Conventional vs. non-conventional formulations for intestinal flora supplementation. *AAPS PharmSciTech, 15*(1), 29–43.

Hakansson, A. P., Orihuela, C. J., & Bogaert, D. (2018). Bacterial-host interactions: Physiology and pathophysiology of respiratory infection. *Physiological Reviews, 98*(2), 781–811.

Haque, S. Z., & Haque, M. (2017). The ecological community of commensal, symbiotic, and pathogenic gastrointestinal microorganisms–an appraisal. *Clinical and Experimental Gastroenterology, 10*, 91.

Hemarajata, P., & Versalovic, J. (2013). Effects of probiotics on gut microbiota: Mechanisms of intestinal immunomodulation and neuromodulation. *Therapeutic Advances in Gastroenterology, 6*(1), 39–51.

Hoare, A., Soto, C., Rojas-Celis, V., & Bravo, D. (2019). Chronic inflammation as a link between periodontitis and carcinogenesis. *Mediators of Inflammation, 2019*, 1.

Hua, S., Marks, E., Schneider, J. J., & Keely, S. (2015). Advances in oral nano-delivery systems for colon targeted drug delivery in inflammatory bowel disease: Selective targeting to diseased versus healthy tissue. *Nanomedicine: Nanotechnology, Biology and Medicine, 11*(5), 1117–1132.

Huang, C., & Shi, G. (2019). Smoking and microbiome in oral, airway, gut and some systemic diseases. *Journal of Translational Medicine, 17*(1), 225.

Jampilek, J., Kos, J., & Kralova, K. (2019). Potential of nanomaterial applications in dietary supplements and foods for special medical purposes. *Nanomaterials, 9*(2), 296.

Javurek, A. B., Suresh, D., Spollen, W. G., Hart, M. L., Hansen, S. A., Ellersieck, M. R., et al. (2017). Gut dysbiosis and neurobehavioral alterations in rats exposed to silver nanoparticles. *Scientific Reports, 7*(1), 2822.

Jiang, Z., Jacob, J. A., Li, J., Wu, X., Wei, G., Vimalanathan, A., et al. (2018). Influence of diet and dietary nanoparticles on gut dysbiosis. *Microbial Pathogenesis, 118*, 61–65.

Jones, S. E., & Versalovic, J. (2009). Probiotic Lactobacillus reuteri biofilms produce antimicrobial and anti-inflammatory factors. *BMC Microbiology, 9*(1), 35.

Jurj, A., Braicu, C., Pop, L. A., Tomuleasa, C., Gherman, C. D., & Berindan-Neagoe, I. (2017). The new era of nanotechnology, an alternative to change cancer treatment. *Drug Design, Development and Therapy, 11*, 2871.

Karavolos, M., & Holban, A. (2016). Nanosized drug delivery systems in gastrointestinal targeting: Interactions with microbiota. *Pharmaceuticals, 9*(4), 62.

Kerry, R. G., Patra, J. K., Gouda, S., Park, Y., Shin, H. S., & Das, G. (2018). Benefaction of probiotics for human health: A review. *Journal of Food and Drug Analysis, 26*(3), 927–939.

Kho, Z. Y., & Lal, S. K. (2018). The human gut microbiome–a potential controller of wellness and disease. *Frontiers in Microbiology, 9*, 1835.

Khushnud, T., & Mousa, S. A. (2013). Potential role of naturally derived polyphenols and their nanotechnology delivery in cancer. *Molecular Biotechnology, 55*(1), 78–86.

Kilian, M., Chapple, I. L. C., Hannig, M., Marsh, P. D., Meuric, V., Pedersen, A. M. L., et al. (2016). The oral microbiome–an update for oral healthcare professionals. *British Dental Journal, 221*(10), 657.

Kimelman, H., & Shemesh, M. (2019). Probiotic bifunctionality of Bacillus subtilis—rescuing lactic acid bacteria from desiccation and antagonizing pathogenic Staphylococcus aureus. *Microorganisms, 7*(10), 407.

Kumar Singh, A., Cabral, C., Kumar, R., Ganguly, R., Kumar Rana, H., Gupta, A., et al. (2019). Beneficial effects of dietary polyphenols on gut microbiota and strategies to improve delivery efficiency. *Nutrients, 11*(9), 2216.

Lamont, R. J., Koo, H., & Hajishengallis, G. (2018). The oral microbiota: Dynamic communities and host interactions. *Nature Reviews Microbiology, 16*(12), 745–759.

Langdon, A., Crook, N., & Dantas, G. (2016). The effects of antibiotics on the microbiome throughout development and alternative approaches for therapeutic modulation. *Genome Medicine, 8*(1), 39.

Lebeaux, D., Ghigo, J. M., & Beloin, C. (2014). Biofilm-related infections: Bridging the gap between clinical management and fundamental aspects of recalcitrance toward antibiotics. *Microbiology and Molecular Biology Reviews, 78*(3), 510–543.

Lebeer, S., Vanderleyden, J., & De Keersmaecker, S. C. (2008). Genes and molecules of lactobacilli supporting probiotic action. *Microbiology and Molecular Biology Reviews, 72*(4), 728–764.

Lee, N. Y., Hsueh, P. R., & Ko, W. C. (2019). Nanoparticles in the treatment of infections caused by multidrug-resistant organisms. *Frontiers in Pharmacology, 10*, 1153.

Li, X., Yeh, Y. C., Giri, K., Mout, R., Landis, R. F., Prakash, Y. S., & Rotello, V. M. (2015). Control of nanoparticle penetration into biofilms through surface design. *Chemical Communications, 51*(2), 282–285.

Lin, L., & Zhang, J. (2017). Role of intestinal microbiota and metabolites on gut homeostasis and human diseases. *BMC Immunology, 18*(1), 2.

Ma, Q., Xing, C., Long, W., Wang, H. Y., Liu, Q., & Wang, R. F. (2019). Impact of microbiota on central nervous system and neurological diseases: The gut-brain axis. *Journal of Neuroinflammation, 16*(1), 53.

Macke, E., Tasiemski, A., Massol, F., Callens, M., & Decaestecker, E. (2017). Life history and eco-evolutionary dynamics in light of the gut microbiota. *Oikos, 126*(4), 508–531.

McFarland, L. V. (2014). Use of probiotics to correct dysbiosis of normal microbiota following disease or disruptive events: A systematic review. *BMJ Open, 4*(8), e005047.

Milani, C., Duranti, S., Bottacini, F., Casey, E., Turroni, F., Mahony, J., et al. (2017). The first microbial colonizers of the human gut: Composition, activities, and health implications of the infant gut microbiota. *Microbiology and Molecular Biology Reviews, 81*(4), e00036–e00017.

Milinčić, D. D., Popović, D. A., Lević, S. M., Kostić, A. Ž., Tešić, Ž. L., Nedović, V. A., & Pešić, M. B. (2019). Application of polyphenol-loaded nanoparticles in food industry. *Nanomaterials, 9*(11), 1629.

Miquel, S., Lagrafeuille, R., Souweine, B., & Forestier, C. (2016). Anti-biofilm activity as a health issue. *Frontiers in Microbiology, 7*, 592.

Mittal, A. K., Kumar, S., & Banerjee, U. C. (2014). Quercetin and gallic acid mediated synthesis of bimetallic (silver and selenium) nanoparticles and their antitumor and antimicrobial potential. *Journal of Colloid and Interface Science, 431*, 194–199.

Mokoena, M. P. (2017). Lactic acid bacteria and their bacteriocins: Classification, biosynthesis and applications against uropathogens: A mini-review. *Molecules, 22*(8), 1255.

Monteagudo-Mera, A., Rastall, R. A., Gibson, G. R., Charalampopoulos, D., & Chatzifragkou, A. (2019). Adhesion mechanisms mediated by probiotics and prebiotics and their potential impact on human health. *Applied Microbiology and Biotechnology, 103*(16), 6463–6472.

Morgillo, F., Dallio, M., Della Corte, C. M., Gravina, A. G., Viscardi, G., Loguercio, C., et al. (2018). Carcinogenesis as a result of multiple inflammatory and oxidative hits: A comprehensive review from tumor microenvironment to gut microbiota. *Neoplasia (New York, NY), 20*(7), 721.

Neuman, H., Forsythe, P., Uzan, A., Avni, O., & Koren, O. (2018). Antibiotics in early life: Dysbiosis and the damage done. *FEMS Microbiology Reviews, 42*(4), 489–499.

Nguyen, S., & Hiorth, M. (2015). Advanced drug delivery systems for local treatment of the oral cavity. *Therapeutic Delivery, 6*(5), 595–608.

Nguyen, H. N., Romero Jovel, S., & Nguyen, T. H. K. (2017). Nanosized minicells generated by lactic acid bacteria for drug delivery. *Journal of Nanomaterials, 2017*.

Ouwehand, A. C., Forssten, S., Hibberd, A. A., Lyra, A., & Stahl, B. (2016). Probiotic approach to prevent antibiotic resistance. *Annals of Medicine, 48*(4), 246–255.

Pagliari, D., Saviano, A., Newton, E. E., Serricchio, M. L., Dal Lago, A. A., Gasbarrini, A., & Cianci, R. (2018). Gut microbiota-immune system crosstalk and pancreatic disorders. *Mediators of Inflammation, 2018*, 7946431.

Parker, A., Fonseca, S., & Carding, S. R. (2020). Gut microbes and metabolites as modulators of blood-brain barrier integrity and brain health. *Gut Microbes, 11*(2), 135–157.

Peters, B. M., Jabra-Rizk, M. A., Graeme, A. O., Costerton, J. W., & Shirtliff, M. E. (2012). Polymicrobial interactions: Impact on pathogenesis and human disease. *Clinical Microbiology Reviews, 25*(1), 193–213.

Plichta, D. R., Graham, D. B., Subramanian, S., & Xavier, R. J. (2019). Therapeutic opportunities in inflammatory bowel disease: Mechanistic dissection of host-microbiome relationships. *Cell, 178*(5), 1041–1056.

Poh, T. Y., Ali, N. A. T. B. M., Mac Aogáin, M., Kathawala, M. H., Setyawati, M. I., Ng, K. W., & Chotirmall, S. H. (2018). Inhaled nanomaterials and the respiratory microbiome: Clinical, immunological and toxicological perspectives. *Particle and Fibre Toxicology, 15*(1), 46.

Postler, T. S., & Ghosh, S. (2017). Understanding the holobiont: How microbial metabolites affect human health and shape the immune system. *Cell Metabolism, 26*(1), 110–130.

Qiu, K., Durham, P. G., & Anselmo, A. C. (2018). Inorganic nanoparticles and the microbiome. *Nano Research, 11*(10), 4936–4954.

Rauta, P. R., Mohanta, Y. K., & Nayak, D. (Eds.). (2019). *Nanotechnology in biology and medicine: Research advancements and future perspectives*. CRC Press. Boca Raton, Florida

Relman, D. A., & Lipsitch, M. (2018). Microbiome as a tool and a target in the effort to address antimicrobial resistance. *Proceedings of the National Academy of Sciences, 115*(51), 12902–12910.

Rinninella, E., Raoul, P., Cintoni, M., Franceschi, F., Miggiano, G. A. D., Gasbarrini, A., & Mele, M. C. (2019). What is the healthy gut microbiota composition? A changing ecosystem across age, environment, diet, and diseases. *Microorganisms, 7*(1), 14.

Rosenberg, E., & Zilber-Rosenberg, I. (2018). The hologenome concept of evolution after 10 years. *Microbiome, 6*(1), 78.

Rosenberg, E., Sharon, G., Atad, I., & Zilber-Rosenberg, I. (2010). The evolution of animals and plants via symbiosis with microorganisms. *Environmental Microbiology Reports, 2*(4), 500–506.

Rosenfeld, C. S. (2017). Gut dysbiosis in animals due to environmental chemical exposures. *Frontiers in Cellular and Infection Microbiology, 7*, 396.

Salas-Jara, M., Ilabaca, A., Vega, M., & García, A. (2016). Biofilm forming Lactobacillus: New challenges for the development of probiotics. *Microorganisms, 4*(3), 35.

Salazar, N., Gueimonde, M., de los Reyes-Gavilan, C. G., & Ruas-Madiedo, P. (2016). Exopolysaccharides produced by lactic acid bacteria and bifidobacteria as fermentable substrates by the intestinal microbiota. *Critical Reviews in Food Science and Nutrition, 56*(9), 1440–1453.

Saus, E., Iraola-Guzmán, S., Willis, J. R., Brunet-Vega, A., & Gabaldón, T. (2019). Microbiome and colorectal cancer: Roles in carcinogenesis and clinical potential. *Molecular Aspects of Medicine, 69*, 93.

Selber-Hnatiw, S., Rukundo, B., Ahmadi, M., Akoubi, H., Al-Bizri, H., Aliu, A. F., et al. (2017). Human gut microbiota: Toward an ecology of disease. *Frontiers in Microbiology, 8*, 1265.

Siemer, S., Hahlbrock, A., Vallet, C., McClements, D. J., Balszuweit, J., Voskuhl, J., et al. (2018). Nanosized food additives impact beneficial and pathogenic bacteria in the human gut: A simulated gastrointestinal study. *NPJ Science of Food, 2*(1), 22.

Simon, J. C., Marchesi, J. R., Mougel, C., & Selosse, M. A. (2019). Host-microbiota interactions: From holobiont theory to analysis. *Microbiome, 7*(1), 5.

Singh, Y., Ahmad, J., Musarrat, J., Ehtesham, N. Z., & Hasnain, S. E. (2013). Emerging importance of holobionts in evolution and in probiotics. *Gut Pathogens, 5*(1), 12.

Sitaraman, R. (2018). Prokaryotic horizontal gene transfer within the human holobiont: Ecological-evolutionary inferences, implications and possibilities. *Microbiome, 6*(1), 163.

Siwek, M., Slawinska, A., Stadnicka, K., Bogucka, J., Dunislawska, A., & Bednarczyk, M. (2018). Prebiotics and synbiotics–in ovo delivery for improved lifespan condition in chicken. *BMC Veterinary Research, 14*(1), 402.

Skillings, D. (2016). Holobionts and the ecology of organisms: Multi-species communities or integrated individuals? *Biology and Philosophy, 31*(6), 875–892.

Stanton, M. M., Park, B. W., Vilela, D., Bente, K., Faivre, D., Sitti, M., & Sánchez, S. (2017). Magnetotactic bacteria powered biohybrids target E. coli biofilms. *ACS Nano, 11*(10), 9968–9978.

Sterlin, D., Fadlallah, J., Slack, E., & Gorochov, G. (2019). The antibody/microbiota interface in health and disease. *Mucosal Immunology, 13*(1), 3–11.

Terpou, A., Papadaki, A., Lappa, I. K., Kachrimanidou, V., Bosnea, L. A., & Kopsahelis, N. (2019). Probiotics in food systems: Significance and emerging strategies towards improved viability and delivery of enhanced beneficial value. *Nutrients, 11*(7), 1591.

Thomas, S., Izard, J., Walsh, E., Batich, K., Chongsathidkiet, P., Clarke, G., et al. (2017). The host microbiome regulates and maintains human health: A primer and perspective for non-microbiologists. *Cancer Research, 77*(8), 1783–1812.

Thursby, E., & Juge, N. (2017). Introduction to the human gut microbiota. *Biochemical Journal, 474*(11), 1823–1836.

Toribio-Mateas, M. (2018). Harnessing the power of microbiome assessment tools as part of neuroprotective nutrition and lifestyle medicine interventions. *Microorganisms, 6*(2), 35.

Tuteja, S., & Ferguson, J. F. (2019). Gut microbiome and response to cardiovascular drugs. *Circulation: Genomic and Precision Medicine, 12*(9), 421–429.

Van Giau, V., Lee, H., An, S. S. A., & Hulme, J. (2019). Recent advances in the treatment of C. difficile using biotherapeutic agents. *Infection and Drug Resistance, 12*, 1597.

Vieco-Saiz, N., Belguesmia, Y., Raspoet, R., Auclair, E., Gancel, F., Kempf, I., & Drider, D. (2019). Benefits and inputs from lactic acid bacteria and their bacteriocins as alternatives to antibiotic growth promoters during food-animal production. *Frontiers in Microbiology, 10*, 57.

Wegh, C. A., Geerlings, S. Y., Knol, J., Roeselers, G., & Belzer, C. (2019). Postbiotics and their potential applications in early life nutrition and beyond. *International Journal of Molecular Sciences, 20*(19), 4673.

Westmeier, D., Hahlbrock, A., Reinhardt, C., Fröhlich-Nowoisky, J., Wessler, S., Vallet, C., et al. (2018). Nanomaterial–microbe cross-talk: Physicochemical principles and (patho) biological consequences. *Chemical Society Reviews, 47*(14), 5312–5337.

Xu, H., Liu, M., Cao, J., Li, X., Fan, D., Xia, Y., et al. (2019). The dynamic interplay between the gut microbiota and autoimmune diseases. *Journal of Immunology Research, 2019*, 1.

Yin, W., Wang, Y., Liu, L., & He, J. (2019). Biofilms: The microbial "protective clothing" in extreme environments. *International Journal of Molecular Sciences, 20*(14), 3423.

Zhang, Z., Tang, H., Chen, P., Xie, H., & Tao, Y. (2019). Demystifying the manipulation of host immunity, metabolism, and extraintestinal tumors by the gut microbiome. *Signal Transduction and Targeted Therapy, 4*(1), 1–34.

Index

J. Bueno, *Preclinical Evaluation of Antimicrobial Nanodrugs*, Nanotechnology in the Life Sciences, https://doi.org/10.1007/978-3-030-43855-5

Printed in the United States
by Baker & Taylor Publisher Services